HEATH
ALGEBRA 2
AN INTEGRATED APPROACH
LARSON, KANOLD, STIFF

LESSON PLANS

McDougal Littell

Evanston, Illinois • Boston • Dallas

International Standard Book Number: 0-395-87963-9

1 2 3 4 5 6 7 8 9 10 HWI 01 00 99 98 97

*Teacher's Name*_____ *Class*_____ *Date*_____ *Room*_____

Goals
    1. Use the real number line to graph and order real numbers.
    2. Use the properties of real numbers.

State/Local Objectives _____

_____

NCTM Curriculum Standards: Problem Solving, Communication, Reasoning, Connections, Structure

✓ **Check items you wish to use for this lesson.**

Introducing the Lesson
____ Teaching Tools: Problem of the Day copymasters page 33, or Teacher's Edition page 2
____ Teaching Tools: Warm-Up Exercises copymasters page 65, or Teacher's Edition page 2

Teaching the Lesson using the following:
____ Color Transparencies page 1
____ Teaching Tools: transparencies page 1

    Notes for substitute teacher _____

_____

_____

Closing the Lesson
____ Communicating about Algebra, Student's Edition page 5
____ Extend Communicating, Teacher's Edition page 5
____ Guided Practice Exercises, Student's Edition page 6

Homework Assignment, pages 6–8
____ Basic/Average: Ex. 7–33 odd, 34–37, 39–55 odd
____ Above Average: Ex. 7–33 odd, 34–37, 39–55 odd
____ Advanced: Ex. 7–33 odd, 34–37, 39–55 odd, 56

Reteaching the Lesson
____ Extra Practice Copymasters page 1
____ Reteaching Copymasters page 1
____ Alternative Assessment, Math Log pages 30–32

Extending the Lesson
____ Enrichment, Teacher's Edition page 8
____ Applications Handbook pages 52, 53
____ Project, Alternative Assessment pages 13–15
____ Technology: Using Calculators and Computers page 1

Notes

_____

_____

_____

Teacher's Name _____ Class _____ Date _____ Room _____

Goals
1. Evaluate an algebraic expression.
2. Use algebraic expressions as models of real-life situations.

State/Local Objectives _____

_____

NCTM Curriculum Standards: Problem Solving, Communication, Connections, Geometry

✓ **Check items you wish to use for this lesson.**

Introducing the Lesson
___ Teaching Tools: Problem of the Day copymasters page 33, or Teacher's Edition page 9
___ Teaching Tools: Warm-Up Exercises copymasters page 65, or Teacher's Edition page 9

Teaching the Lesson using the following:
___ Color Transparencies pages 1, 2
___ Teaching Tools: transparencies pages 2, 3
___ Technology for Example 2, Teacher's Edition page 10

    Notes for substitute teacher _____

_____

_____

Closing the Lesson
___ Communicating about Algebra, Student's Edition page 11
___ Extend Communicating, Teacher's Edition page 11
___ Guided Practice Exercises, Student's Edition page 12

Homework Assignment, pages 12–14
___ Basic/Average: Ex. 7–45 odd, 49–52, 54, 55–59 odd, 64–66
___ Above Average: Ex. 7–49 odd, 49–52, 54, 55–63 odd, 64–67
___ Advanced: Ex. 11–47 odd, 49–52, 53–63 odd, 64–67

Reteaching the Lesson
___ Extra Practice Copymasters page 2
___ Reteaching Copymasters page 2

Extending the Lesson
___ Enrichment, Teacher's Edition pages 14–15
___ Applications Handbook pages 18, 19
___ Alternative Assessment page 14

Notes

_____

_____

_____

*Teacher's Name*_____ *Class*_____ *Date*_____ *Room*_____

Goals   1. Solve a linear equation.
              2. Use linear equations to answer questions about real life.

State/Local Objectives _____

_____

NCTM Curriculum Standards: Problem Solving, Communication, Connections

✓ **Check items you wish to use for this lesson.**

Introducing the Lesson
___ Teaching Tools: Problem of the Day copymasters page 33, or Teacher's Edition page 18
___ Teaching Tools: Warm-Up Exercises copymasters page 65, or Teacher's Edition page 18

   Notes for substitute teacher _____

_____

_____

Closing the Lesson
___ Communicating about Algebra, Student's Edition page 20
___ Guided Practice Exercises, Student's Edition page 21

Homework Assignment, pages 21–23
___ Basic/Average: Ex. 5–43 odd, 49, 63
___ Above Average: Ex. 5–43 odd, 45–49, 63–65
___ Advanced: Ex. 5–43 odd, 44–50, 63–65

Reteaching the Lesson
___ Extra Practice Copymasters page 3
___ Reteaching Copymasters page 3

Extending the Lesson
___ Enrichment, Teacher's Edition page 22

Notes

_____

_____

_____

Teacher's Name _____ Class _____ Date _____ Room _____

Goal   Use problem-solving strategies to solve real-life problems.

State/Local Objectives _____

_____

NCTM Curriculum Standards: Problem Solving, Communication, Connections

✓ **Check items you wish to use for this lesson.**

Introducing the Lesson
___ Teaching Tools: Problem of the Day copymasters page 34, or Teacher's Edition page 24
___ Teaching Tools: Warm-Up Exercises copymasters page 66, or Teacher's Edition page 24

Teaching the Lesson using the following:
___ Color Transparencies page 3
___ Teaching Tools: transparencies page 4

   Notes for substitute teacher _____

_____

_____

Closing the Lesson
___ Communicating about Algebra, Student's Edition page 26
___ Guided Practice Exercises, Student's Edition page 27

Homework Assignment, pages 27–29
___ Basic/Average: Ex. 5–8, 15–16, 18–24, 30–31
___ Above Average: Ex. 5–8, 13–24, 31
___ Advanced: Ex. 5–12, 15–24, 31

Reteaching the Lesson
___ Extra Practice Copymasters page 4
___ Reteaching Copymasters page 4

Extending the Lesson
___ Enrichment, Teacher's Edition page 29

Notes

_____

_____

_____

Teacher's Name _____ Class _____ Date _____ Room _____

Goal   Solve a literal equation for a given variable and evaluate it for specified values of the other variables.

State/Local Objectives _____

_____

NCTM Curriculum Standards: Problem Solving, Communication, Connections, Geometry

✓ **Check items you wish to use for this lesson.**

Introducing the Lesson
___ Teaching Tools: Problem of the Day copymasters page 34, or Teacher's Edition page 31
___ Teaching Tools: Warm-Up Exercises copymasters page 66, or Teacher's Edition page 31

Teaching the Lesson using the following:
___ Color Transparencies page 4
___ Extension for Example 2, Teacher's Edition page 32

Notes for substitute teacher _____

_____

_____

Closing the Lesson
___ Communicating about Algebra, Student's Edition page 33
___ Extend Communicating, Teacher's Edition page 33
___ Guided Practice Exercises, Student's Edition page 34

Homework Assignment, pages 34–36
___ Basic/Average: Ex. 5–10, 11–23 odd, 25–27, 21–39 odd
___ Above Average: Ex. 5–10, 11–23 odd, 24–27, 31–39, 40–41
___ Advanced: Ex. 5–12, 13–23 odd, 24–27, 31–39 odd, 40–41

Reteaching the Lesson
___ Extra Practice Copymasters page 5
___ Reteaching Copymasters page 5

Extending the Lesson
___ Enrichment, Teacher's Edition page 36

Notes

_____

_____

_____

Teacher's Name_____ Class_____ Date_____ Room_____

Goals
1. Solve simple and compound inequalities.
2. Use inequalities to solve real-life problems.

State/Local Objectives _____

_____

NCTM Curriculum Standards: Problem Solving, Communication, Connections

✓ **Check items you wish to use for this lesson.**

Introducing the Lesson
____ Teaching Tools: Problem of the Day copymasters page 35, or Teacher's Edition page 37
____ Teaching Tools: Warm-Up Exercises copymasters page 66, or Teacher's Edition page 37

Teaching the Lesson using the following:
____ Teaching Tools: transparencies page 1

Notes for substitute teacher _____

_____

_____

Closing the Lesson
____ Communicating about Algebra, Student's Edition page 40
____ Guided Practice Exercises, Student's Edition page 40

Homework Assignment, pages 41–43
____ Basic/Average: Ex. 7–14, 17–37 odd, 39–43, 46–56
____ Above Average: Ex. 9–14, 19–37 odd, 39–45, 52–57
____ Advanced: Ex. 11–14, 21–37 odd, 39–45, 52–58

Reteaching the Lesson
____ Extra Practice Copymasters page 6
____ Reteaching Copymasters page 6

Extending the Lesson
____ Enrichment, Teacher's Edition page 43
____ Applications Handbook pages 8, 9, 54, 55, 56
____ Cultural Diversity Extensions page 27

Notes

_____

_____

_____

*Teacher's Name*_____ *Class*_____ *Date*_____ *Room*_____

Goals   1. Solve absolute value equations and inequalities.
              2. Use absolute value inequalities in real-life settings.

State/Local Objectives _____

_____

NCTM Curriculum Standards: Problem Solving, Communication, Connections

✓ **Check items you wish to use for this lesson.**

Introducing the Lesson
____ Teaching Tools: Problem of the Day copymasters page 35, or Teacher's Edition page 44
____ Teaching Tools: Warm-Up Exercises copymasters page 67, or Teacher's Edition page 44

Teaching the Lesson using the following:
____ Color Transparencies page 5
____ Extension for Example 4, Teacher's Edition page 45
____ Teaching Tools: transparcncies page 1

     Notes for substitute teacher _____

_____

_____

Closing the Lesson
____ Communicating about Algebra, Student's Edition page 46
____ Guided Practice Exercises, Student's Edition page 47

Homework Assignment, pages 47–49
____ Basic/Average: Ex. 5–33 odd, 35–36, 41–48, 58–60
____ Above Average: Ex. 7–35 odd, 37–48, 58–62
____ Advanced: Ex. 9–37, 38–48, 59–62

Reteaching the Lesson
____ Extra Practice Copymasters page 7
____ Reteaching Copymasters page 7

Extending the Lesson
____ Cooperative Learning, Alternative Assessment page 14
____ Enrichment, Teacher's Edition pages 48–49

Notes

_____

_____

_____

Teacher's Name _____ Class _____ Date _____ Room _____

Goal   Organize data by using tables and graphs.

State/Local Objectives _____

_____

NCTM Curriculum Standards: Problem Solving, Communication, Connections, Discrete Math

✓ **Check items you wish to use for this lesson.**

Introducing the Lesson
____ Teaching Tools: Problem of the Day copymasters page 35, or Teacher's Edition page 50
____ Teaching Tools: Warm-Up Exercises copymasters page 67, or Teacher's Edition page 50

Teaching the Lesson using the following:
____ Color Transparencies pages 6, 7, 8
____ Teaching Tools: transparencies pages 5, 6

Notes for substitute teacher _____

_____

_____

Closing the Lesson
____ Communicating about Algebra, Student's Edition page 52
____ Extend Communicating, Teacher's Edition page 53
____ Guided Practice Exercises, Student's Edition page 53

Homework Assignment, pages 53–55
____ Basic/Average: Ex. 5, 7–15, 17–23 odd
____ Above Average: Ex. 5–15, 17–23 odd
____ Advanced: Ex. 5–16, 21–23

Reteaching the Lesson
____ Extra Practice Copymasters page 8
____ Reteaching Copymasters page 8

Extending the Lesson
____ Cooperative Learning, Alternative Assessment pages 14–15
____ Enrichment, Teacher's Edition page 55
____ Technology: Using Calculators and Computers page 4

Notes

_____

_____

_____

© D. C. Heath and Company

Teacher's Name _____ Class _____ Date _____ Room _____

Goals
    1. Sketch a graph using a table of values.
    2. Identify equations whose graphs are horizontal or vertical lines.

State/Local Objectives _____

_____

NCTM Curriculum Standards: Problem Solving, Communication, Connections

✓ **Check items you wish to use for this lesson.**

Introducing the Lesson
___ Teaching Tools: Problem of the Day copymasters page 35, or Teacher's Edition page 64
___ Teaching Tools: Warm-Up Exercises copymasters page 67, or Teacher's Edition page 64

Teaching the Lesson using the following:
___ Extension for Example 4, Teacher's Edition page 65
___ Color Transparencies page 10
___ Teaching Tools: transparencies pages 7, 8, 9

    Notes for substitute teacher _____

_____

_____

Closing the Lesson
___ Communicating about Algebra, Student's Edition page 66
___ Extend Communicating, Teacher's Edition page 66
___ Guided Practice Exercises, Student's Edition page 67

Homework Assignment, pages 67–69
___ Basic/Average: Ex. 7–12, 13–17 odd, 19–24, 25–45 odd, 47–48, 55–60
___ Above Average: Ex. 7–24, 25–47 odd, 48, 57–61
___ Advanced: Ex. 13–24, 25–47 odd, 48, 59–62

Reteaching the Lesson
___ Extra Practice Copymasters page 9
___ Reteaching Copymasters page 9
___ Alternative Assessment: Math Log pages 33–35

Extending the Lesson
___ Enrichment, Teacher's Edition page 69
___ Cooperative Learning, Alternative Assessment pages 15–16
___ Research Project, Alternative Assessment pages 15–16

Notes

_____

_____

_____

Teacher's Name _____ Class _____ Date _____ Room _____

Goals   1. Find the slope of a line and identify parallel and perpendicular lines from their slopes.
2. Interpret slope as a rate of change.

State/Local Objectives _____

_____

NCTM Curriculum Standards: Problem Solving, Communication, Connections, Geometry (algebraic), Calculus
concepts

✓ **Check items you wish to use for this lesson.**

Introducing the Lesson
___ Teaching Tools: Problem of the Day copymasters page 36, or Teacher's Edition page 70
___ Teaching Tools: Warm-Up Exercises copymasters page 68, or Teacher's Edition page 70

Teaching the Lesson using the following:
___ Common-Error Alert, Teacher's Edition page 71
___ Color Transparencies page 11
___ Teaching Tools: transparencies pages 7, 8, 9

Notes for substitute teacher _____

_____

_____

Closing the Lesson
___ Communicating about Algebra, Student's Edition page 73
___ Extend Communicating, Teacher's Edition page 73
___ Guided Practice Exercises, Student's Edition page 74

Homework Assignment, pages 74–76
___ Basic/Average: Ex. 5–8, 11–21 odd, 27–30, 33–43 odd, 44, 55–59 odd
___ Above Average: Ex. 5–10, 11–23 odd, 27–30, 33–43 odd, 44, 57–62
___ Advanced: Ex. 5–8, 9–25 odd, 27–30, 31–41 odd, 42–44, 57–62

Reteaching the Lesson
___ Extra Practice Copymasters page 10
___ Reteaching Copymasters page 10

Extending the Lesson
___ Enrichment, Teacher's Edition page 76
___ Cooperative Learning, Alternative Assessment page 16
___ Demonstration, Alternative Assessment page 16
___ Technology: Using Calculators and Computers page 6

Notes

_____

_____

Teacher's Name_____ Class_____ Date_____ Room_____

Goals   1. Use intercepts to sketch a quick graph of a line.
            2. Use the slope-intercept form of a line to sketch a quick graph.

State/Local Objectives _____

_____

NCTM Curriculum Standards: Problem Solving, Communication, Connections

✓ **Check items you wish to use for this lesson.**

Introducing the Lesson
____ Teaching Tools: Problem of the Day copymasters page 36, or Teacher's Edition page 78
____ Teaching Tools: Warm-Up Exercises copymasters page 68, or Teacher's Edition page 78

Teaching the Lesson using the following:
____ Teaching Tools: transparencies pages 2, 3, 7, 8, 9

     Notes for substitute teacher _____

_____

_____

Closing the Lesson
____ Communicating about Algebra, Student's Edition page 80
____ Extend Communicating, Teacher's Edition page 80
____ Guided Practice Exercises, Student's Edition page 81

Homework Assignment, pages 81–83
____ Basic/Average: Ex. 11, 13–18, 19–39 odd, 43–46, 47–59 odd, 67, 69, 70
____ Above Average: Ex. 11–18, 23–39 odd, 43–47, 67–71
____ Advanced: Ex. 13–18, 23–26, 31–43 odd, 44–46, 67–72

Reteaching the Lesson
____ Extra Practice Copymasters page 11
____ Reteaching Copymasters page 11

Extending the Lesson
____ Enrichment, Teacher's Edition page 83
____ Cultural Diversity Extensions page 28

Notes

_____

_____

_____

2 days for Basic Course, 1 day for Full Course

Teacher's Name _____ Class _____ Date _____ Room _____

Goals
    1. Write equations of lines.
    2. Use equations of lines to solve real-life problems.

State/Local Objectives _____

_____

NCTM Curriculum Standards: Problem Solving, Communication, Connections, Calculus concepts

✓ **Check items you wish to use for this lesson.**

Introducing the Lesson
___ Teaching Tools: Problem of the Day copymasters page 36, or Teacher's Edition page 86
___ Teaching Tools: Warm-Up Exercises copymasters page 69, or Teacher's Edition page 86

Teaching the Lesson using the following:
___ Color Transparencies pages 11, 12
___ Teaching Tools: transparencies pages 7, 8, 9

    Notes for substitute teacher _____

_____

_____

Closing the Lesson
___ Communicating about Algebra, Student's Edition page 89
___ Extend Communicating, Teacher's Edition page 89
___ Guided Practice Exercises, Student's Edition page 90

Homework Assignment, pages 90–92
___ Basic/Average: Ex. 5–25 odd, 28–29, 33–37 odd, 38–41, 43–49 odd, 52
___ Above Average: Ex. 9–25 odd, 27–30, 38–41, 43–51 odd, 52
___ Advanced: Ex. 9–25 odd, 26–30, 38–41, 48–52

Reteaching the Lesson
___ Extra Practice Copymasters page 12
___ Reteaching Copymasters page 12

Extending the Lesson
___ Technology, Teacher's Edition page 91
___ Enrichment, Teacher's Edition page 92
___ Applications Handbook page 41

Notes

_____

_____

_____

Teacher's Name _____ Class _____ Date _____ Room _____

Goals 1. Graph a linear inequality in two variables.
2. Use a linear inequality in two variables to model real-life situations.

State/Local Objectives _____

_____

NCTM Curriculum Standards: Problem Solving, Communication, Connections, Geometry

✓ **Check items you wish to use for this lesson.**

Introducing the Lesson
____ Teaching Tools: Problem of the Day copymasters page 36, or Teacher's Edition page 94
____ Teaching Tools: Warm-Up Exercises copymasters page 69, or Teacher's Edition page 94

Teaching the Lesson using the following:
____ Teaching Tools: transparencies pages 7, 8, 10, 11

Notes for substitute teacher _____

_____

_____

Closing the Lesson
____ Communicating about Algebra, Student's Edition page 96
____ Extend Communicating, Teacher's Edition page 96
____ Guided Practice Exercises, Student's Edition page 97

Homework Assignment, pages 97–99
____ Basic/Average: Ex. 5–15 odd, 17–28, 29–39 odd, 40–42, 51–55, 59, 61
____ Above Average: Ex. 7–15 odd, 17–28, 29–39 odd, 40–42, 51–55, 59, 61
____ Advanced: Ex. 13–23 odd, 25–28, 29–37 odd, 38–42, 51–56, 59–63

Reteaching the Lesson
____ Extra Practice Copymasters page 13
____ Reteaching Copymasters page 13

Extending the Lesson
____ Technology, Teacher's Edition page 99
____ Enrichment, Teacher's Edition page 100
____ Cooperative Learning, Alternative Assessment page 16

Notes

_____

_____

_____

Teacher's Name _____ *Class* _____ *Date* _____ *Room* _____

Goals
1. Graph absolute value equations.
2. Use graphs of absolute value equations to answer questions about real-life situations.

State/Local Objectives _____

_____

NCTM Curriculum Standards: Problem Solving, Communication, Reasoning, Connections

✓ **Check items you wish to use for this lesson.**

Introducing the Lesson
___ Teaching Tools: Problem of the Day copymasters page 37, or Teacher's Edition page 101
___ Teaching Tools: Warm-Up Exercises copymasters page 70, or Teacher's Edition page 101

Teaching the Lesson using the following:
___ Connections for Example 3, Teacher's Edition page 102
___ Color Transparencies page 13
___ Teaching Tools: transparencies pages 7, 8, 12

   Notes for substitute teacher _____

_____

_____

Closing the Lesson
___ Communicating about Algebra, Student's Edition page 103
___ Guided Practice Exercises, Student's Edition page 104

Homework Assignment, pages 104–106
___ Basic/Average: Ex. 5–8, 9–19 odd, 21–24, 27–39 odd, 44, 56–58
___ Above Average: Ex. 5–8, 9–19 odd, 21–24, 27–41 odd, 45–48, 55–58
___ Advanced: Ex. 5–8, 17–24, 27–41 odd, 45–48, 56–60

Reteaching the Lesson
___ Extra Practice Copymasters page 14
___ Reteaching Copymasters page 14

Extending the Lesson
___ Enrichment, Teacher's Edition page 106
___ Technology: Using Calculators and Computers pages 9, 12

Notes

_____

_____

_____

Lesson Plan 2.7
pages 107–113

Teacher's Name _____ *Class* _____ *Date* _____ *Room* _____

Goals
1. Fit a line to a set of data and write an equation for the line.
2. Identify whether a set of data shows positive or negative correlation, or no correlation.

State/Local Objectives _____

_____

NCTM Curriculum Standards: Problem Solving, Communication, Connections, Statistics, Discrete Math

✓ **Check items you wish to use for this lesson.**

Introducing the Lesson
___ Teaching Tools: Problem of the Day copymasters page 37, or Teacher's Edition page 107
___ Teaching Tools: Warm-Up Exercises copymasters page 70, or Teacher's Edition page 107

Teaching the Lesson using the following:
___ Color Transparencies pages 14, 15
___ Teaching Tools: transparencies pages 5, 7, 8, 9

Notes for substitute teacher _____

_____

_____

Closing the Lesson
___ Communicating about Algebra, Student's Edition page 110
___ Extend Communicating, Teacher's Edition page 110
___ Guided Practice Exercises, Student's Edition page 111

Homework Assignment, pages 111–113
___ Basic/Average: Ex. 5–8, 9–13 odd, 16–19, 27–34
___ Above Average: Ex. 5–8, 9–15 odd, 16–19, 27–36
___ Advanced: Ex. 5–8, 9–15 odd, 16–18, 27–36

Reteaching the Lesson
___ Extra Practice Copymasters page 15
___ Reteaching Copymasters page 15

Extending the Lesson
___ Enrichment, Teacher's Edition page 113
___ Cooperative Learning, Alternative Assessment page 17
___ Technology: Using Calculators and Computers page 14

Notes

_____

_____

_____

Teacher's Name_____ Class_____ Date_____ Room_____

Goals
    1. Graph and solve a system of linear equations.
    2. Use a system of linear equations to answer questions about a real-life situation.

State/Local Objectives _____

_____

NCTM Curriculum Standards: Problem Solving, Communication, Reasoning, Connections

✓ **Check items you wish to use for this lesson.**

Introducing the Lesson
____ Teaching Tools: Problem of the Day copymasters page 37, or Teacher's Edition page 122
____ Teaching Tools: Warm-Up Exercises copymasters page 70, or Teacher's Edition page 122

Teaching the Lesson using the following:
____ Color Transparencies pages 18, 19
____ Teaching Tools: transparencies pages 2, 3, 8, 9

    Notes for substitute teacher _____

_____

_____

Closing the Lesson
____ Communicating about Algebra, Student's Edition page 124
____ Guided Practice Exercises, Student's Edition page 125

Homework Assignment, pages 125–127
____ Basic/Average: Ex. 7–17 odd, 23–35 odd, 38–40, 42–45, 48, 51
____ Above Average: Ex. 7–19 odd, 25–37 odd, 38–40, 42–45, 56–58
____ Advanced: Ex. 7–19 odd, 25–37 odd, 38–44, 57–58

Reteaching the Lesson
____ Extra Practice Copymasters page 16
____ Reteaching Copymasters page 16
____ Alternative Assessment: Math Log pages 36–38

Extending the Lesson
____ Technology, Teacher's Edition page 126
____ Enrichment, Teacher's Edition page 127
____ Alternative Assessment page 17
____ Technology: Using Calculators and Computers page 17

Notes

_____

_____

_____

Teacher's Name _____ *Class* _____ *Date* _____ *Room* _____

Goals   1. Use algebraic methods to solve a linear system.
2. Use a linear system to answer questions about a real-life situation.

State/Local Objectives _____

_____

NCTM Curriculum Standards: Problem Solving, Communication, Reasoning, Connections

✓ **Check items you wish to use for this lesson.**

Introducing the Lesson
___ Teaching Tools: Problem of the Day copymasters page 37, or Teacher's Edition page 130
___ Teaching Tools: Warm-Up Exercises copymasters page 71, or Teacher's Edition page 130

Teaching the Lesson using the following:
___ Color Transparencies pages 20, 21
___ Teaching Tools: transparencies pages 7, 8, 9

Notes for substitute teacher _____

_____

_____

Closing the Lesson
___ Communicating about Algebra, Student's Edition page 133
___ Extend Communicating, Teacher's Edition page 133
___ Guided Practice Exercises, Student's Edition page 134

Homework Assignment, pages 134–136
___ Basic/Average: Ex. 9–19 odd, 23–27 odd, 29–31, 37–45, 51–52, 54–56
___ Above Average: Ex. 9–19 odd, 21–29 odd, 30–32, 48–56
___ Advanced: Ex. 7–21 odd, 27, 29, 30–32, 49–56

Reteaching the Lesson
___ Cooperative Learning, Alternative Assessment page 18
___ Extra Practice Copymasters page 17
___ Reteaching Copymasters page 17

Extending the Lesson
___ Enrichment, Teacher's Edition page 135

Notes

_____

_____

_____

Teacher's Name _____ Class _____ Date _____ Room _____

Goal   Write and use linear systems to model real-life situations.

State/Local Objectives _____

_____

NCTM Curriculum Standards: Problem Solving, Communication, Connections, Statistics, Discrete Math

✓ **Check items you wish to use for this lesson.**

Introducing the Lesson
____ Teaching Tools: Problem of the Day copymasters page 38, or Teacher's Edition page 138
____ Teaching Tools: Warm-Up Exercises copymasters page 71, or Teacher's Edition page 138

Teaching the Lesson using the following:
____ Color Transparencies pages 21, 22
____ Teaching Tools: transparencies pages 4, 7, 8, 9

   Notes for substitute teacher _____

_____

_____

Closing the Lesson
____ Communicating about Algebra, Student's Edition page 140
____ Guided Practice Exercises, Student's Edition page 141

Homework Assignment, pages 141–143
____ Basic/Average: Ex. 5–9 odd, 12, 13, 19–23 odd, 24, 25
____ Above Average: Ex. 5–11 odd, 12–15, 23–25
____ Advanced: Ex. 5–11 odd, 17–23 odd, 24, 25

Reteaching the Lesson
____ Extra Practice Copymasters page 18
____ Reteaching Copymasters page 18

Extending the Lesson
____ Enrichment, Teacher's Edition page 144
____ Cultural Diversity Extensions page 29

Notes

_____

_____

_____

Teacher's Name_____ Class_____ Date_____ Room_____

Goals   1. Graph a system of linear inequalities to find the solution to the system.
              2. Use a system of linear inequalities to model a real-life situation.

State/Local Objectives _____

_____

NCTM Curriculum Standards: Problem Solving, Communication, Reasoning, Connections, Discrete Math

✓ **Check items you wish to use for this lesson.**

Introducing the Lesson
____ Teaching Tools: Problem of the Day copymasters page 38, or Teacher's Edition page 145
____ Teaching Tools: Warm-Up Exercises copymasters page 71, or Teacher's Edition page 145

Teaching the Lesson using the following:
____ Color Transparencies pages 23, 24
____ Teaching Tools: transparencies pages 8, 10, 11

Notes for substitute teacher _____

_____

_____

Closing the Lesson
____ Communicating about Algebra, Student's Edition page 147
____ Extend Communicating, Teacher's Edition page 147
____ Guided Practice Exercises, Student's Edition page 148

Homework Assignment, pages 148–150
____ Basic/Average: Ex. 7–12, 13–27 odd, 31–35 odd, 39–53 odd
____ Above Average: Ex. 7–12, 15–37 odd, 43–53 odd, 54
____ Advanced: Ex. 7–12, 17–37 odd, 38, 49–54

Reteaching the Lesson
____ Extra Practice Copymasters page 19
____ Reteaching Copymasters page 19

Extending the Lesson
____ Technology, Teacher's Edition page 148
____ Technology, Teacher's Edition page 149
____ Enrichment, Teacher's Edition page 150
____ Cooperative Learning, Alternative Assessment page 18

Notes

_____

_____

_____

Teacher's Name _____ Class _____ Date _____ Room _____

Goals
1. Solve a linear programming problem.
2. Use linear programming to answer questions about real-life situations.

State/Local Objectives _____

_____

NCTM Curriculum Standards: Problem Solving, Communication, Connections, Discrete Math

✓ **Check items you wish to use for this lesson.**

Introducing the Lesson
____ Teaching Tools: Problem of the Day copymasters page 39, or Teacher's Edition page 151
____ Teaching Tools: Warm-Up Exercises copymasters page 71, or Teacher's Edition page 151

Teaching the Lesson using the following:
____ Connections for Example 1, Teacher's Edition page 152
____ Color Transparencies page 25
____ Teaching Tools: transparencies pages 8, 10, 11

Notes for substitute teacher _____

_____

_____

Closing the Lesson
____ Communicating about Algebra, Student's Edition page 153
____ Extend Communicating, Teacher's Edition page 154
____ Guided Practice Exercises, Student's Edition page 154

Homework Assignment, pages 154–156
____ Basic/Average: Ex. 5–21 odd, 22, 23, 29–31
____ Above Average: Ex. 7–21 odd, 22–25, 28–32
____ Advanced: Ex. 7–21 odd, 22–24, 28–33

Reteaching the Lesson
____ Extra Practice Copymasters page 20
____ Reteaching Copymasters page 20

Extending the Lesson
____ Enrichment, Teacher's Edition page 156

Notes

_____

_____

_____

Teacher's Name_____ Class_____ Date_____ Room_____

Goals
1. Solve a system of linear equations in three variables.
2. Use a system in three variables to answer questions about real-life situations.

State/Local Objectives _____

_____

NCTM Curriculum Standards: Problem Solving, Communication, Connections

✓ **Check items you wish to use for this lesson.**

Introducing the Lesson
___ Teaching Tools: Problem of the Day copymasters page 39, or Teacher's Edition page 157
___ Teaching Tools: Warm-Up Exercises copymasters page 72, or Teacher's Edition page 157

Notes for substitute teacher _____

_____

_____

Closing the Lesson
___ Communicating about Algebra, Student's Edition page 160
___ Extend Communicating, Teacher's Edition page 161
___ Guided Practice Exercises, Student's Edition page 161

Homework Assignment, pages 161–163
___ Basic/Average: Ex. 9–21 odd, 27–35 odd, 41
___ Above Average: Ex. 9–21 odd, 25–31 odd, 41, 42
___ Advanced: Ex. 9–21 odd, 25–31 odd, 41–43

Reteaching the Lesson
___ Extra Practice Copymasters page 21
___ Reteaching Copymasters page 21

Extending the Lesson
___ Enrichment, Teacher's Edition page 163

Notes

_____

_____

_____

Teacher's Name _____ Class _____ Date _____ Room _____

Goals
1. Organize data into matrices and work with them, using addition, subtraction, and scalar multiplication.
2. Use matrices in real-life settings.

State/Local Objectives _____

_____

NCTM Curriculum Standards: Problem Solving, Communication, Connections, Discrete Math

✓ **Check items you wish to use for this lesson.**

Introducing the Lesson
____ Teaching Tools: Problem of the Day copymasters page 39, or Teacher's Edition page 174
____ Teaching Tools: Warm-Up Exercises copymasters page 72, or Teacher's Edition page 174

Notes for substitute teacher _____

_____

_____

Closing the Lesson
____ Communicating about Algebra, Student's Edition page 176
____ Guided Practice Exercises, Student's Edition page 177

Homework Assignment, pages 177–179
____ Basic/Average: Ex. 7–31 odd, 32–37, 39–45 odd
____ Above Average: Ex. 7–31 odd, 37, 39–45 odd, 46
____ Advanced: Ex. 7–33 odd, 34–37, 46

Reteaching the Lesson
____ Extra Practice Copymasters page 22
____ Reteaching Copymasters page 22
____ Alternative Assessment: Math Log pages 39–41

Extending the Lesson
____ Enrichment, Teacher's Edition page 179
____ Cooperative Learning: Alternative Assessment pages 18–19

Notes

_____

_____

_____

*Teacher's Name*_____ *Class*_____ *Date*_____ *Room*_____

Goals   1. Multiply two matrices.
              2. Use matrix multiplication to answer questions about real-life situations.

State/Local Objectives _____

_____

NCTM Curriculum Standards: Problem Solving, Communication, Connections, Discrete Math, Structure

✓ **Check items you wish to use for this lesson.**

Introducing the Lesson
____ Teaching Tools: Problem of the Day copymasters page 39, or Teacher's Edition page 180
____ Teaching Tools: Warm-Up Exercises copymasters page 73, or Teacher's Edition page 180

Teaching the Lesson using the following:
____ Extension for Example 3, Teacher's Edition page 182

      Notes for substitute teacher ____ _____

_____

_____

Closing the Lesson
____ Communicating about Algebra, Student's Edition page 182
____ Extend Communicating, Teacher's Edition page 183
____ Guided Practice Exercises, Student's Edition page 183

Homework Assignment, pages 183–185
____ Basic/Average: Ex. 5–12, 13–31 odd, 34–43
____ Above Average: Ex. 5–12, 13–33 odd, 35–46
____ Advanced: Ex. 5–33 odd, 38–47

Reteaching the Lesson
____ Extra Practice Copymasters page 23
____ Reteaching Copymasters page 23

Extending the Lesson
____ Enrichment, Teacher's Edition page 185
____ Technology, Teacher's Edition page 185

Notes

_____

_____

_____

Teacher's Name_____ Class_____ Date_____ Room_____

Goals   1. Evaluate the determinant of a 2×2 or a 3×3 matrix.
              2. Use determinants to solve real-life problems.

State/Local Objectives _____

_____

NCTM Curriculum Standards: Problem Solving, Communication, Connections, Geometry (algebraic), Discrete Math

✓ **Check items you wish to use for this lesson.**

Introducing the Lesson
____ Teaching Tools: Problem of the Day copymasters page 40, or Teacher's Edition page 187
____ Teaching Tools: Warm-Up Exercises copymasters page 73, or Teacher's Edition page 187

Teaching the Lesson using the following:
____ Color Transparencies pages 26, 27
____ Teaching Tools: transparencies pages 2, 3

      Notes for substitute teacher _____

_____

_____

Closing the Lesson
____ Communicating about Algebra, Student's Edition page 190
____ Guided Practice Exercises, Student's Edition page 191

Homework Assignment, pages 191–193
____ Basic/Average: Ex. 7–29 odd, 33, 37, 38, 40, 41–46
____ Above Average: Ex. 7–31 odd, 37, 39, 40, 43–49
____ Advanced: Ex. 7–31 odd, 37, 39, 40, 43–49

Reteaching the Lesson
____ Extra Practice Copymasters page 24
____ Reteaching Copymasters page 24

Extending the Lesson
____ Technology, Teacher's Edition page 191
____ Enrichment, Teacher's Edition page 193
____ Cultural Diversity Extensions page 30

Notes

_____

_____

_____

Teacher's Name_____ Class_____ Date_____ Room_____

Goals   1. Find and use the inverse of a 2×2 matrix.
            2. Use inverse matrices in real-life settings.

State/Local Objectives _____

_____

NCTM Curriculum Standards: Problem Solving, Communication, Connections, Discrete Math, Structure

✓ **Check items you wish to use for this lesson.**

Introducing the Lesson
____ Teaching Tools: Problem of the Day copymasters page 40, or Teacher's Edition page 196
____ Teaching Tools: Warm-Up Exercises copymasters page 74, or Teacher's Edition page 196

Teaching the Lesson using the following:
____ Common-Error Alert, Teacher's Edition page 197

   Notes for substitute teacher _____

   _____

   _____

Closing the Lesson
____ Communicating about Algebra, Student's Edition page 199
____ Guided Practice Exercises, Student's Edition page 200

Homework Assignment, pages 200–202
____ Basic/Average: Ex. 7–29 odd, 35–39 odd, 40–42, 43–49 odd
____ Above Average: Ex. 7–29 odd, 35, 37, 39–43, 47–52
____ Advanced: Ex. 7–29 odd, 35, 37, 39–44, 47–52

Reteaching the Lesson
____ Extra Practice Copymasters page 25
____ Reteaching Copymasters page 25

Extending the Lesson
____ Technology, Teacher's Edition page 201
____ Enrichment, Teacher's Edition page 202
____ Cooperative Learning, Alternative Assessment page 19
____ Technology: Using Calculators and Computers pages 19, 21

Notes

_____

_____

_____

Lesson Plan 4.5
pages 204–209

Teacher's Name _____ Class _____ Date _____ Room _____

Goals
1. Solve systems of linear equations using inverse matrices.
2. Use systems of linear equations to solve real-life problems.

State/Local Objectives _____

_____

NCTM Curriculum Standards: Problem Solving, Communication, Connections, Discrete Math, Structure, Technology

✓ **Check items you wish to use for this lesson.**

Introducing the Lesson
___ Teaching Tools: Problem of the Day copymasters page 40, or Teacher's Edition page 204
___ Teaching Tools: Warm-Up Exercises copymasters page 74, or Teacher's Edition page 204

Notes for substitute teacher _____

_____

_____

Closing the Lesson
___ Communicating about Algebra, Student's Edition page 206
___ Guided Practice Exercises, Student's Edition page 207

Homework Assignment, pages 207–209
___ Basic/Average: Ex. 5–21 odd, 27–31 odd, 32, 33–39 odd
___ Above Average: Ex. 5–31 odd, 32, 33–39 odd, 40
___ Advanced: Ex. 9–27 odd, 29–32, 35, 39, 40

Reteaching the Lesson
___ Extra Practice Copymasters page 26
___ Reteaching Copymasters page 26

Extending the Lesson
___ Technology, Teacher's Edition page 208
___ Technology, Teacher's Edition page 209
___ Enrichment, Teacher's Edition page 209

Notes

_____

_____

_____

26 *Algebra 2*

Teacher's Name_____ Class_____ Date_____ Room_____

Goals
    1. Use an augmented matrix to solve a system of linear equations.
    2. Use a system of linear equations to solve real-life problems.

State/Local Objectives _____

_____

NCTM Curriculum Standards: Problem Solving, Communication, Connections, Discrete Math, Structure

✓ **Check items you wish to use for this lesson.**

Introducing the Lesson
___ Teaching Tools: Problem of the Day copymasters page 40, or Teacher's Edition page 210
___ Teaching Tools: Warm-Up Exercises copymasters page 75, or Teacher's Edition page 210

    Notes for substitute teacher _____

_____

_____

Closing the Lesson
___ Communicating about Algebra, Student's Edition page 212
___ Guided Practice Exercises, Student's Edition page 213

Homework Assignment, pages 213–215
___ Basic/Average: Ex. 7, 15–19 odd, 21–24, 34–35
___ Above Average: Ex. 7, 15–19 odd, 20–23, 26, 35–37
___ Advanced: Ex. 7, 15–19 odd, 20–23, 35–37

Reteaching the Lesson
___ Extra Practice Copymasters page 27
___ Reteaching Copymasters page 27

Extending the Lesson
___ Technology, Teacher's Edition page 214
___ Enrichment, Teacher's Edition page 215
___ Technology: Using Calculators and Computers page 23

Notes

_____

_____

_____

Teacher's Name _____ Class _____ Date _____ Room _____

Goals
1. Use Cramer's Rule to solve a system of linear equations.
2. Use linear equations to solve real-life problems.

State/Local Objectives _____

_____

NCTM Curriculum Standards: Problem Solving, Communication, Connections, Discrete Math

✓ **Check items you wish to use for this lesson.**

Introducing the Lesson
___ Teaching Tools: Problem of the Day copymasters page 41, or Teacher's Edition page 216
___ Teaching Tools: Warm-Up Exercises copymasters page 75, or Teacher's Edition page 216

Teaching the Lesson using the following:
___ Extension for Examples 1–2, Teacher's Edition page 217
___ Technology for Examples 1–2, Teacher's Edition page 217
___ Color Transparencies pages 28–29

Notes for substitute teacher _____

_____

_____

Closing the Lesson
___ Communicating about Algebra, Student's Edition page 218
___ Extend Communicating, Teacher's Edition page 218
___ Guided Practice Exercises, Student's Edition page 219

Homework Assignment, pages 219–221
___ Basic/Average: Ex. 5–25 odd, 26–31, 33–40, 43
___ Above Average: Ex. 5–25 odd, 26–31, 33–35, 40–43
___ Advanced: Ex. 9–25 odd, 26–35, 40–41

Reteaching the Lesson
___ Extra Practice Copymasters page 28
___ Reteaching Copymasters page 28

Extending the Lesson
___ Enrichment, Teacher's Edition page 221

Notes

_____

_____

_____

Teacher's Name _____ Class _____ Date _____ Room _____

Goals
1. Solve a quadratic equation by finding square roots.
2. Use quadratic equations as models of real-life situations.

State/Local Objectives _____

_____

NCTM Curriculum Standards: Problem Solving, Communication, Connections, Geometry (synthetic)

✓ **Check items you wish to use for this lesson.**

Introducing the Lesson
____ Teaching Tools: Problem of the Day copymasters page 41, or Teacher's Edition page 230
____ Teaching Tools: Warm-Up Exercises copymasters page 76, or Teacher's Edition page 230

Teaching the Lesson using the following:
____ Common-Error Alert, Teacher's Edition page 230
____ Problem Solving Activity for Example 2, Teacher's Edition page 231

Notes for substitute teacher _____

_____

_____

Closing the Lesson
____ Communicating about Algebra, Student's Edition page 232
____ Guided Practice Exercises, Student's Edition page 233

Homework Assignment, pages 233–235
____ Basic/Average: Ex. 5–9 odd, 17–27 odd, 29–33, 35–39 odd, 49–53
____ Above Average: Ex. 11–27 odd, 29–35, 49–53
____ Advanced: Ex. 11–29 odd, 30–36, 49–53

Reteaching the Lesson
____ Extra Practice Copymasters page 29
____ Reteaching Copymasters page 29
____ Alternative Assessment: Math Log pages 42–44

Extending the Lesson
____ Technology, Teacher's Edition page 234
____ Enrichment, Teacher's Edition page 235
____ Applications Handbook pages 42–51
____ Demonstration, Alternative Assessment page 19

Notes

_____

_____

Teacher's Name_____ Class_____ Date_____ Room_____

Goals
    1. Graph a quadratic equation.
    2. Use the graph of a quadratic equation in a real-life setting.

State/Local Objectives _____

_____

NCTM Curriculum Standards: Problem Solving, Communication, Connections, Geometry (algebraic)

✓ **Check items you wish to use for this lesson.**

Introducing the Lesson
____ Teaching Tools: Problem of the Day copymasters page 41, or Teacher's Edition page 236
____ Teaching Tools: Warm-Up Exercises copymasters page 76, or Teacher's Edition page 236

Teaching the Lesson using the following:
____ Color Transparencies pages 32, 33
____ Extension for Example 3, Teacher's Edition page 237
____ Teaching Tools: transparencies pages 2, 3, 7, 8, 13

    Notes for substitute teacher _____

_____

_____

Closing the Lesson
____ Communicating about Algebra, Student's Edition page 238
____ Guided Practice Exercises, Student's Edition page 239

Homework Assignment, pages 239–241
____ Basic/Average: Ex. 5–19 odd, 23–28, 35–41 odd, 51–59 odd, 60–61
____ Above Average: Ex. 7–21 odd, 23–28, 33–41 odd, 44–46, 57–62
____ Advanced: Ex. 9–21 odd, 23–28, 33–39 odd, 43–46, 59–62

Reteaching the Lesson
____ Extra Practice Copymasters page 30
____ Reteaching Copymasters page 30

Extending the Lesson
____ Enrichment, Teacher's Edition page 241
____ Cultural Diversity Extensions page 31
____ Technology: Using Calculators and Computers pages 25, 28

Notes

_____

_____

_____

Teacher's Name _____ Class _____ Date _____ Room _____

Goals
    1. Solve a quadratic equation by completing the square.
    2. Use *completing the square* to solve real-life problems.

State/Local Objectives _____

_____

NCTM Curriculum Standards: Problem Solving, Communication, Connections, Geometry (algebraic)

✓ **Check items you wish to use for this lesson.**

Introducing the Lesson
____ Teaching Tools: Problem of the Day copymasters page 41, or Teacher's Edition page 245
____ Teaching Tools: Warm-Up Exercises copymasters page 77, or Teacher's Edition page 245

Teaching the Lesson using the following:
____ Teaching Tools: transparencies pages 7, 8, 13

    Notes for substitute teacher _____

_____

_____

Closing the Lesson
____ Communicating about Algebra, Student's Edition page 247
____ Guided Practice Exercises, Student's Edition page 248

Homework Assignment, pages 248–250
____ Basic/Average: Ex. 9–19 odd, 23–29 odd, 33–45 odd
____ Above Average: Ex. 11–21 odd, 27–35 odd, 45–51 odd
____ Advanced: Ex. 13–27 odd, 28–32, 50–55

Reteaching the Lesson
____ Extra Practice Copymasters page 31
____ Reteaching Copymasters page 31

Extending the Lesson
____ Technology, Teacher's Edition page 249
____ Enrichment, Teacher's Edition page 250
____ Technology: Using Calculators and Computers page 31

Notes

_____

_____

_____

Teacher's Name _____ Class _____ Date _____ Room _____

Goals   1. Use the quadratic formula to solve quadratic equations.
2. Use quadratic models in real-life settings.

State/Local Objectives _____

_____

NCTM Curriculum Standards: Problem Solving, Communication, Connections

✓ **Check items you wish to use for this lesson.**

Introducing the Lesson
___ Teaching Tools: Problem of the Day copymasters page 42, or Teacher's Edition page 252
___ Teaching Tools: Warm-Up Exercises copymasters page 77, or Teacher's Edition page 252

Teaching the Lesson using the following:
___ Extension for Example 3, Teacher's Edition page 253
___ Common-Error Alert, Teacher's Edition page 255
___ Color Transparencies pages 33, 34
___ Teaching Tools: transparencies pages 7, 8, 13

Notes for substitute teacher _____

_____

_____

Closing the Lesson
___ Communicating about Algebra, Student's Edition page 254
___ Guided Practice Exercises, Student's Edition page 255

Homework Assignment, pages 255–257
___ Basic/Average: Ex. 7–27 odd, 29–41 odd, 49–53 odd, 57–60
___ Above Average: Ex. 7–27 odd, 29–41 odd, 44–48, 54–60
___ Advanced: Ex. 7–27 odd, 29–41 odd, 44–48, 53–60

Reteaching the Lesson
___ Extra Practice Copymasters page 32
___ Reteaching Copymasters page 32

Extending the Lesson
___ Enrichment, Teacher's Edition page 256
___ Technology, Teacher's Edition page 256
___ Applications Handbook pages 46–48
___ Technology: Using Calculators and Computers page 33

Notes

_____

_____

Teacher's Name _____ Class _____ Date _____ Room _____

Goals   1. Identify, add, subtract, and multiply imaginary and complex numbers.
              2. Plot complex numbers in the complex plane.

State/Local Objectives _____

_____

NCTM Curriculum Standards: Problem Solving, Communication, Connections, Structure

✓ **Check items you wish to use for this lesson.**

Introducing the Lesson
___ Teaching Tools: Problem of the Day copymasters page 42, or Teacher's Edition page 258
___ Teaching Tools: Warm-Up Exercises copymasters page 77, or Teacher's Edition page 258

Teaching the Lesson using the following:
___ Extension for Example 3, Teacher's Edition page 259
___ Color Transparencies pages 34, 35
___ Teaching Tools: transparencies pages 7, 8

   Notes for substitute teacher _____

_____

_____

Closing the Lesson
___ Communicating about Algebra, Student's Edition page 261
___ Extend Communicating: Teacher's Edition page 261
___ Guided Practice Exercises, Student's Edition page 262

Homework Assignment, pages 262–264
___ Basic/Average: Ex. 13–23 odd, 29–39 odd, 49–57 odd, 65, 69–72, 85–88
___ Above Average: Ex. 13–23 odd, 31–41 odd, 51–61 odd, 66, 69–72, 85–88
___ Advanced: Ex. 15–25 odd, 31–41 odd, 51–61 odd, 68–72, 85–88

Reteaching the Lesson
___ Extra Practice Copymasters page 33
___ Reteaching Copymasters page 33

Extending the Lesson
___ Enrichment, Teacher's Edition page 263
___ Cooperative Learning, Alternative Assessment page 19
___ Applications Handbook pages 26–28

Notes

_____

_____

_____

Teacher's Name_____ Class_____ Date_____ Room_____

Goals
    1. Solve any quadratic equation (even ones that have complex number solutions).
    2. Use complex numbers and programmable calculators to solve real-life problems.

State/Local Objectives _____

_____

NCTM Curriculum Standards: Problem Solving, Communication, Connections, Structure, Technology

✓ **Check items you wish to use for this lesson.**

Introducing the Lesson
____ Teaching Tools: Problem of the Day copymasters page 42, or Teacher's Edition page 264
____ Teaching Tools: Warm-Up Exercises copymasters page 78, or Teacher's Edition page 264

    Notes for substitute teacher _____

_____

_____

Closing the Lesson
____ Communicating about Algebra, Student's Edition page 267
____ Extend Communicating: Cooperative Learning, Teacher's Edition page 267
____ Guided Practice Exercises, Student's Edition page 268

Homework Assignment, pages 268–270
____ Basic/Average: Ex. 9–19 odd, 23–27 odd, 29–33 odd, 34, 39, 44, 45, 49
____ Above Average: Ex. 9–19 odd, 23–27 odd, 29–34, 39, 40, 47, 51
____ Advanced: Ex. 9–19 odd, 23–27 odd, 29–34, 37, 43, 49, 51

Reteaching the Lesson
____ Extra Practice Copymasters page 34
____ Reteaching Copymasters page 34

Extending the Lesson
____ Technology, Teacher's Edition page 266
____ Enrichment, Teacher's Edition pages 269–270
____ Project: Alternative Assessment page 20

Notes

_____

_____

_____

Teacher's Name _____ *Class* _____ *Date* _____ *Room* _____

Goals
1. Sketch the graph of a quadratic inequality.
2. Use quadratic inequalities in real-life modeling.

State/Local Objectives _____

_____

NCTM Curriculum Standards: Problem Solving, Communication, Connections

✓ **Check items you wish to use for this lesson.**

Introducing the Lesson
___ Teaching Tools: Problem of the Day copymasters page 42, or Teacher's Edition page 271
___ Teaching Tools: Warm-Up Exercises copymasters page 78, or Teacher's Edition page 271

Teaching the Lesson using the following:
___ Extension for Example 3, Teacher's Edition page 272
___ Color Transparencies page 36
___ Teaching Tools: transparencies pages 7, 8, 13

Notes for substitute teacher _____

_____

_____

Closing the Lesson
___ Communicating about Algebra, Student's Edition page 273
___ Guided Practice Exercises, Student's Edition page 274

Homework Assignment, pages 274–276
___ Basic/Average: Ex. 11–16, 23–29 odd, 33–37 odd, 47–51 odd
___ Above Average: Ex. 7, 11–16, 21–29 odd, 33–36, 54–56
___ Advanced: Ex. 9, 11–16, 21–29 odd, 32–36, 55–56

Reteaching the Lesson
___ Extra Practice Copymasters page 35
___ Reteaching Copymasters page 35

Extending the Lesson
___ Technology, Teacher's Edition page 275
___ Enrichment, Teacher's Edition page 276

Notes

_____

_____

_____

Teacher's Name _____ Class _____ Date _____ Room _____

Goals   1. Identify a relation and a function.
              2. Identify real-life relations that are functions.

State/Local Objectives _____

_____

NCTM Curriculum Standards: Problem Solving, Communication, Reasoning, Connections, Functions, Geometry

✓ **Check items you wish to use for this lesson.**

Introducing the Lesson
___ Teaching Tools: Problem of the Day copymasters page 43, or Teacher's Edition page 284
___ Teaching Tools: Warm-Up Exercises copymasters page 79, or Teacher's Edition page 284

Teaching the Lesson using the following:
___ Technology for Example 3, Teacher's Edition page 285
___ Teaching Tools: transparencies pages 7, 8

   Notes for substitute teacher _____

_____

_____

Closing the Lesson
___ Communicating about Algebra, Student's Edition page 287
___ Extend Communicating, Teacher's Edition page 287
___ Guided Practice Exercises, Student's Edition page 288

Homework Assignment, pages 288–290
___ Basic/Average: Ex. 5–19 odd, 21, 22, 23–35 odd, 37–39, 45
___ Above Average: Ex. 5–21 odd, 22, 23–35 odd, 36–38, 45, 47
___ Advanced: Ex. 5–21 odd, 23–35 odd, 36–38, 45, 47

Reteaching the Lesson
___ Extra Practice Copymasters page 36
___ Reteaching Copymasters page 36
___ Alternative Assessment: Math Log pages 45–47

Extending the Lesson
___ Enrichment, Teacher's Edition page 290
___ Cooperative Learning, Alternative Assessment page 20
___ Cultural Diversity Extensions page 32

Notes

_____

_____

36 *Algebra 2*

Teacher's Name _____ *Class* _____ *Date* _____ *Room* _____

Goals
1. Perform operations with functions.
2. Use function operations in real-life situations.

State/Local Objectives _____

_____

NCTM Curriculum Standards: Problem Solving, Communication, Reasoning, Connections, Functions, Structure

✓ **Check items you wish to use for this lesson.**

Introducing the Lesson
___ Teaching Tools: Problem of the Day copymasters page 43, or Teacher's Edition page 291
___ Teaching Tools: Warm-Up Exercises copymasters page 79, or Teacher's Edition page 291

Teaching the Lesson using the following:
___ Extension for Examples 2–3, Teacher's Edition page 292
___ Color Transparencies page 37

Notes for substitute teacher _____

_____

_____

Closing the Lesson
___ Communicating about Algebra, Student's Edition page 293
___ Extend Communicating, Teacher's Edition page 293
___ Guided Practice Exercises, Student's Edition page 294

Homework Assignment, pages 294–296
___ Basic/Average: Ex. 5–12, 15–23 odd, 31, 33, 34, 37–41, 53–58
___ Above Average: Ex. 5–12, 15–23 odd, 33, 34, 35–43, 55–59
___ Advanced: Ex. 5–12, 15–23, 33, 35–43, 56–60

Reteaching the Lesson
___ Extra Practice Copymasters page 37
___ Reteaching Copymasters page 37

Extending the Lesson
___ Enrichment, Teacher's Edition page 296
___ Technology, Teacher's Edition page 296
___ Cooperative Learning, Alternative Assessment page 20
___ Technology: Using Calculators and Computers page 35

Notes

_____

_____

_____

Teacher's Name_____ Class_____ Date_____ Room_____

Goals
1. Identify inverse relations and inverse functions and verify that two functions are inverses of each other.
2. Use inverse functions in real-life situations.

State/Local Objectives _____

_____

NCTM Curriculum Standards: Problem Solving, Communication, Reasoning, Connections, Functions

✓ **Check items you wish to use for this lesson.**

Introducing the Lesson
____ Teaching Tools: Problem of the Day copymasters page 43, or Teacher's Edition page 298
____ Teaching Tools: Warm-Up Exercises copymasters page 79, or Teacher's Edition page 298

Teaching the Lesson using the following:
____ Alternative Approach for Example 1, Teacher's Edition page 299
____ Teaching Tools: transparencies pages 7, 8, 9, 13

Notes for substitute teacher _____

_____

_____

Closing the Lesson
____ Communicating about Algebra, Student's Edition page 301
____ Extend Communicating, Teacher's Edition page 301
____ Guided Practice Exercises, Student's Edition page 302

Homework Assignment, pages 302–304
____ Basic/Average: Ex. 10–14, 21–37 odd, 41–45 odd, 46, 51–54, 57–58
____ Above Average: Ex. 10–14, 21–37 odd, 41–46, 51–54, 58–60
____ Advanced: Ex. 10–14, 21–37, 41–46, 51–54, 59–62

Reteaching the Lesson
____ Extra Practice Copymasters page 38
____ Reteaching Copymasters page 38

Extending the Lesson
____ Enrichment, Teacher's Edition page 304
____ Applications Handbook page 54

Notes

_____

_____

_____

Teacher's Name _____ Class _____ Date _____ Room _____

Goals
1. Use special functions, such as compound functions and step functions.
2. Use these functions in real-life situations.

State/Local Objectives _____

_____

NCTM Curriculum Standards: Problem Solving, Communication, Connections, Functions

✓ **Check items you wish to use for this lesson.**

Introducing the Lesson
___ Teaching Tools: Problem of the Day copymasters page 44, or Teacher's Edition page 306
___ Teaching Tools: Warm-Up Exercises copymasters page 79, or Teacher's Edition page 306

Teaching the Lesson using the following:
___ Alternative Approach for Example 1, Teacher's Edition page 307
___ Color Transparencies page 37
___ Teaching Tools: transparencies pages 2, 3, 7, 8, 12, 13

Notes for substitute teacher _____

_____

_____

Closing the Lesson
___ Communicating about Algebra, Student's Edition page 308
___ Extend Communicating, Teacher's Edition page 309
___ Guided Practice Exercises, Student's Edition page 309

Homework Assignment, pages 309–311
___ Basic/Average: Ex. 5–15 odd, 19–22, 27–33 odd, 34, 37–38, 45–50
___ Above Average: Ex. 5–15 odd, 19–22, 25–31 odd, 33–38, 43, 51
___ Advanced: Ex. 5–17 odd, 19–23, 31–38, 50–51

Reteaching the Lesson
___ Extra Practice Copymasters page 39
___ Reteaching Copymasters page 39

Extending the Lesson
___ Technology, Teacher's Edition page 310
___ Enrichment, Teacher's Edition page 311

Notes

_____

_____

_____

Teacher's Name_____ Class_____ Date_____ Room_____

Goals
1. Use translations and reflections to sketch the graph of a function.
2. Use transformations of graphs of functions in real-life settings.

State/Local Objectives _____

_____

NCTM Curriculum Standards: Problem Solving, Communication, Connections, Functions, Geometry (algebraic)

✓ **Check items you wish to use for this lesson.**

Introducing the Lesson
____ Teaching Tools: Problem of the Day copymasters page 44, or Teacher's Edition page 314
____ Teaching Tools: Warm-Up Exercises copymasters page 80, or Teacher's Edition page 314

Teaching the Lesson using the following:
____ Alternative Approach, Teacher's Edition page 316
____ Teaching Tools: transparencies pages 7, 8, 12, 13

Notes for substitute teacher _____

_____

_____

Closing the Lesson
____ Communicating about Algebra, Student's Edition page 317
____ Extend Communicating: Cooperative Learning, Teacher's Edition page 318
____ Guided Practice Exercises, Student's Edition page 318

Homework Assignment, pages 318–320
____ Basic/Average: Ex. 13–19 odd, 21–24, 33–41 odd, 43–49, 57–58
____ Above Average: Ex. 13–19 odd, 21–24, 31–43 odd, 42–50, 57–58
____ Advanced: Ex. 13–19 odd, 21–24, 31–43 odd, 42–50, 57–60

Reteaching the Lesson
____ Extra Practice Copymasters page 40
____ Reteaching Copymasters page 40

Extending the Lesson
____ Research, Teacher's Edition page 319
____ Technology, Teacher's Edition page 319
____ Enrichment, Teacher's Edition page 320
____ Technology: Using Calculators and Computers page 37

Notes

_____

_____

_____

Teacher's Name_____ Class_____ Date_____ Room_____

Goals   1. Identify and classify recursive functions.
              2. Use recursive functions as real-life models.

State/Local Objectives _____

_____

NCTM Curriculum Standards: Problem Solving, Communication, Reasoning, Connections, Functions,
                           Discrete Math, Structure

✓ **Check items you wish to use for this lesson.**

Introducing the Lesson
___ Teaching Tools: Problem of the Day copymasters page 44, or Teacher's Edition page 321
___ Teaching Tools: Warm-Up Exercises copymasters page 80, or Teacher's Edition page 321

Teaching the Lesson using the following:
___ Color Transparencies page 36

   Notes for substitute teacher _____

_____

_____

Closing the Lesson
___ Communicating about Algebra, Student's Edition page 323
___ Extend Communicating, Teacher's Edition page 323
___ Guided Practice Exercises, Student's Edition page 324

Homework Assignment, pages 324–326
___ Basic/Average: Ex. 11–21 odd, 29–37 odd, 41–47 odd, 55, 58, 60, 71
___ Above Average: Ex. 11–21 odd, 27–37 odd, 45–55 odd, 57–60, 72–74
___ Advanced: Ex. 11–21 odd, 27–37 odd, 45–55 odd, 57–60, 72–74

Reteaching the Lesson
___ Extra Practice Copymasters page 41
___ Reteaching Copymasters page 41

Extending the Lesson
___ Enrichment, Teacher's Edition page 326
___ Writing, Teacher's Edition page 326

Notes

_____

_____

_____

Teacher's Name _____ Class _____ Date _____ Room _____

Goals
    1. Find the mean, median, mode, and quartiles of a set of numbers.
    2. Use measures of central tendency in real-life situations.

State/Local Objectives _____

_____

NCTM Curriculum Standards: Problem Solving, Communication, Reasoning, Connections, Functions, Discrete Math, Structure

✓ **Check items you wish to use for this lesson.**

Introducing the Lesson
___ Teaching Tools: Problem of the Day copymasters page 45, or Teacher's Edition page 327
___ Teaching Tools: Warm-Up Exercises copymasters page 81, or Teacher's Edition page 327

Teaching the Lesson using the following:
___ Color Transparencies pages 38, 39
___ Teaching Tools: transparencies pages 1, 5

Notes for substitute teacher _____

_____

_____

Closing the Lesson
___ Communicating about Algebra, Student's Edition page 329
___ Guided Practice Exercises, Student's Edition page 330

Homework Assignment, pages 330–332
___ Basic/Average: Ex. 5–9, 10–12, 13–14, 17–20
___ Above Average: Ex. 5–20
___ Advanced: Ex. 5–20

Reteaching the Lesson
___ Extra Practice Copymasters page 42
___ Reteaching Copymasters page 42

Extending the Lesson
___ Enrichment, Teacher's Edition page 332
___ Cooperative Learning, Alternative Assessment page 21
___ Technology: Using Calculators and Computers page 40

Notes

_____

_____

_____

Teacher's Name _____ Class _____ Date _____ Room _____

Goals
1. Use properties of exponents to evaluate and simplify exponential expressions.
2. Use powers as models in real-life problems.

State/Local Objectives _____

_____

NCTM Curriculum Standards: Problem Solving, Communication, Reasoning, Connections, Geometry, Structure

✓ **Check items you wish to use for this lesson.**

Introducing the Lesson
____ Teaching Tools: Problem of the Day copymasters page 45, or Teacher's Edition page 346
____ Teaching Tools: Warm-Up Exercises copymasters page 81, or Teacher's Edition page 346

Teaching the Lesson using the following:
____ Common-Error Alert, Teacher's Edition page 349
____ Teaching Tools: transparencies pages 7, 8

    Notes for substitute teacher _____

_____

_____

Closing the Lesson
____ Communicating about Algebra, Student's Edition page 348
____ Guided Practice Exercises, Student's Edition page 349

Homework Assignment, pages 349–351
____ Basic/Average: Ex. 9–27 odd, 43–49 odd, 50, 51, 54–56, 60–62
____ Above Average: Ex. 9–31 odd, 37–43 odd, 48–56, 60–62
____ Advanced: Ex. 9–31 odd, 35–43 odd, 48–56, 60–63

Reteaching the Lesson
____ Extra Practice Copymasters page 43
____ Reteaching Copymasters page 43
____ Alternative Assessment: Math Log pages 48–50

Extending the Lesson
____ Technology, Teacher's Edition page 349
____ Technology, Teacher's Edition page 350
____ Enrichment, Teacher's Edition page 351
____ Applications Handbook page 4

Notes

_____

_____

_____

Teacher's Name _____ Class _____ Date _____ Room _____

Goals
1. Use the compound interest formula and the exponential growth and decay formulas.
2. Solve real-life problems modeled by exponential growth formulas.

State/Local Objectives _____

_____

NCTM Curriculum Standards: Problem Solving, Communication, Connections

✓ **Check items you wish to use for this lesson.**

Introducing the Lesson
____ Teaching Tools: Problem of the Day copymasters page 45, or Teacher's Edition page 352
____ Teaching Tools: Warm-Up Exercises copymasters page 81, or Teacher's Edition page 352

Teaching the Lesson using the following:
____ Technology for Example 4, Teacher's Edition page 353
____ Extension for Example 5: Chemistry, Teacher's Edition page 355
____ Color Transparencies page 41
____ Teaching Tools: transparencies pages 5, 7, 8, 14, 15

____ Notes for substitute teacher _____

_____

_____

Closing the Lesson
____ Communicating about Algebra, Student's Edition page 355
____ Guided Practice Exercises, Student's Edition page 356

Homework Assignment, pages 356–358
____ Basic/Average: Ex. 7–17 odd, 19–22, 27–31 odd, 35–39
____ Above Average: Ex. 7–17 odd, 19–22, 25–31 odd, 32, 37–39
____ Advanced: Ex. 11–19 odd, 20–23, 27–32, 37–39

Reteaching the Lesson
____ Extra Practice Copymasters page 44
____ Reteaching Copymasters page 44

Extending the Lesson
____ Technology, Teacher's Edition page 357
____ Enrichment, Teacher's Edition page 358
____ Cultural Diversity Extensions page 33
____ Research Project, Alternative Assessment page 21
____ Technology: Using Calculators and Computers pages 42, 45

Notes

_____

_____

_____

Lesson Plan 7.3
pages 360–366

2 days for Basic Course, 1 day for Full Course

Teacher's Name_____ Class_____ Date_____ Room_____

Goals
1. Evaluate nth roots of real numbers using radical notation and rational exponent notation.
2. Use nth roots to solve real-life problems.

State/Local Objectives _____

_____

NCTM Curriculum Standards: Problem Solving, Communication, Connections, Technology

✓ **Check items you wish to use for this lesson.**

Introducing the Lesson
___ Teaching Tools: Problem of the Day copymasters page 46, or Teacher's Edition page 360
___ Teaching Tools: Warm-Up Exercises copymasters page 82, or Teacher's Edition page 360

Teaching the Lesson using the following:
___ Extension for Example 5, Teacher's Edition page 361
___ Technology for Example 3, Teacher's Edition page 361

Notes for substitute teacher _____

_____

_____

Closing the Lesson
___ Communicating about Algebra, Student's Edition page 363
___ Extend Communicating, Teacher's Edition page 363
___ Guided Practice Exercises, Student's Edition page 364

Homework Assignment, pages 364–366
___ Basic/Average: Ex. 7–31 odd, 39–49 odd, 59–67 odd, 78
___ Above Average: Ex. 7–33 odd, 39–49 odd, 57–65 odd, 66–68, 77–82
___ Advanced: Ex. 9–35 odd, 41–45 odd, 57–65 odd, 66–68, 77–82

Reteaching the Lesson
___ Extra Practice Copymasters page 45
___ Reteaching Copymasters page 45

Extending the Lesson
___ Technology, Teacher's Edition page 364
___ Technology, Teacher's Edition page 365
___ Enrichment, Teacher's Edition page 366

Notes

_____

_____

_____

2 days for Basic Course, 1 day for Full Course

Teacher's Name_____ Class_____ Date_____ Room_____

Goals
   1. Use properties of roots to evaluate and simplify expressions containing radicals and rational exponents.
   2. Use properties of roots to solve real-life problems.

State/Local Objectives _____

_____

NCTM Curriculum Standards: Problem Solving, Communication, Connections

✓ **Check items you wish to use for this lesson.**

Introducing the Lesson
____ Teaching Tools: Problem of the Day copymasters page 46, or Teacher's Edition page 368
____ Teaching Tools: Warm-Up Exercises copymasters page 82, or Teacher's Edition page 368

Teaching the Lesson using the following:
____ Color Transparencies pages 41, 42

   Notes for substitute teacher _____

_____

_____

Closing the Lesson
____ Communicating about Algebra, Student's Edition page 370
____ Guided Practice Exercises, Student's Edition page 371

Homework Assignment, pages 371–373
____ Basic/Average: Ex. 9–25 odd, 39–47 odd, 63–65, 67, 69–72, 75–89 odd
____ Above Average: Ex. 13–27 odd, 39–49 odd, 61–65, 67–74, 75–89 odd, 90–92
____ Advanced: Ex. 13–29 odd, 41–65 odd, 67–74, 75–89 odd, 90–92

Reteaching the Lesson
____ Extra Practice Copymasters page 46
____ Reteaching Copymasters page 46

Extending the Lesson
____ Enrichment, Teacher's Edition page 373
____ Applications Handbook page 56

Notes

_____

_____

_____

Teacher's Name_____ Class_____ Date_____ Room_____

Goals
1. Solve equations that have radicals and rational exponents.
2. Use radical equations to solve real-life problems.

State/Local Objectives _____

_____

NCTM Curriculum Standards: Problem Solving, Communication, Connections, Geometry (algebraic)

✓ **Check items you wish to use for this lesson.**

Introducing the Lesson
___ Teaching Tools: Problem of the Day copymasters page 46, or Teacher's Edition page 374
___ Teaching Tools: Warm-Up Exercises copymasters page 82, or Teacher's Edition page 374

Teaching the Lesson using the following:
___ Common-Error Alert, Teacher's Edition page 374
___ Extension for Examples 1–2, Teacher's Edition page 375
___ Extension for Example 4, Teacher's Edition page 375
___ Color Transparencies page 42
___ Teaching Tools: transparencies pages 7, 8

Notes for substitute teacher _____

_____

_____

Closing the Lesson
___ Communicating about Algebra, Student's Edition page 377
___ Guided Practice Exercises, Student's Edition page 378

Homework Assignment, pages 378–380
___ Basic/Average: Ex. 9–27 odd, 45, 47, 51–54, 55–63 odd, 64
___ Above Average: Ex. 9–27 odd, 41–47, 51–54, 55–63 odd, 64
___ Advanced: Ex. 9–29 odd, 37–47 odd, 50–54, 63–65

Reteaching the Lesson
___ Extra Practice Copymasters page 47
___ Reteaching Copymasters page 47

Extending the Lesson
___ Enrichment, Teacher's Edition page 380
___ Technology: Using Calculators and Computers page 48

Notes

_____

_____

_____

2 days for Basic Course, 1 day for Full Course

Teacher's Name_____ Class_____ Date_____ Room_____

Goals
1. Graph square root and cube root functions.
2. Use square root and cube root functions to solve real-life problems.

State/Local Objectives _____

_____

NCTM Curriculum Standards: Problem Solving, Communication, Connections, Functions, Geometry (algebraic)

✓ **Check items you wish to use for this lesson.**

Introducing the Lesson
___ Teaching Tools: Problem of the Day copymasters page 47, or Teacher's Edition page 381
___ Teaching Tools: Warm-Up Exercises copymasters page 83, or Teacher's Edition page 381

Teaching the Lesson using the following:
___ Color Transparencies page 43
___ Teaching Tools: transparencies pages 2, 3, 7, 8, 13, 16

Notes for substitute teacher _____

_____

_____

Closing the Lesson
___ Communicating about Algebra, Student's Edition page 383
___ Guided Practice Exercises, Student's Edition page 384

Homework Assignment, pages 384–387
___ Basic/Average: Ex. 9, 11–16, 21–31 odd, 35–42, 51–56
___ Above Average: Ex. 9, 11–16, 19–31 odd, 35–42, 51–56, 59–60
___ Advanced: Ex. 9, 11–16, 19–33 odd, 35–40, 43, 44, 51–60

Reteaching the Lesson
___ Extra Practice Copymasters page 48
___ Reteaching Copymasters page 48

Extending the Lesson
___ Technology, Teacher's Edition page 385
___ Enrichment, Teacher's Edition page 386
___ Technology, Teacher's Edition page 386
___ Alternative Assessment page 22

Notes

_____

_____

_____

© D.C. Heath and Company

Teacher's Name _____ Class _____ Date _____ Room _____

Goals
    1. Graph exponential functions and evaluate exponential expressions.
    2. Use exponential functions as models for real-life situations.

State/Local Objectives _____

_____

NCTM Curriculum Standards: Problem Solving, Communication, Connections, Functions, Geometry (algebraic),
                               Structure

✓ **Check items you wish to use for this lesson.**

Introducing the Lesson
____ Teaching Tools: Problem of the Day copymasters page 47, or Teacher's Edition page 398
____ Teaching Tools: Warm-Up Exercises copymasters page 83, or Teacher's Edition page 398

Teaching the Lesson using the following:
____ Technology for Example 4, Teacher's Edition page 400
____ Color Transparencies page 44
____ Teaching Tools: transparencies pages 7, 8, 14, 15

    Notes for substitute teacher _____

_____

_____

Closing the Lesson
____ Communicating about Algebra, Student's Edition page 401
____ Guided Practice Exercises, Student's Edition page 402

Homework Assignment, pages 402–404
____ Basic/Average: Ex. 7–11 odd, 13–20, 27–39 odd, 46, 47–51 odd, 65–66
____ Above Average: Ex. 7–11 odd, 13–20, 27–39 odd, 45–52, 53–60, 65–66
____ Advanced: Ex. 7–11 odd, 13–20, 21–29 odd, 35–51 odd, 52, 65–72

Reteaching the Lesson
____ Extra Practice Copymasters page 49
____ Reteaching Copymasters page 49
____ Alternative Assessment: Math Log pages 51–53

Extending the Lesson
____ Technology, Teacher's Edition page 401
____ Technology, Teacher's Edition page 403
____ Enrichment, Teacher's Edition page 404

Notes

_____

_____

Teacher's Name_____ Class_____ Date_____ Room_____

Goals
    1. Evaluate logarithmic expressions and graph logarithmic functions.
    2. Use logarithms in real-life situations.

State/Local Objectives _____

_____

NCTM Curriculum Standards: Problem Solving, Communication, Connections, Functions, Structure

✓ **Check items you wish to use for this lesson.**

Introducing the Lesson
___ Teaching Tools: Problem of the Day copymasters page 47, or Teacher's Edition page 405
___ Teaching Tools: Warm-Up Exercises copymasters page 83, or Teacher's Edition page 405

Teaching the Lesson using the following:
___ Teaching Tools: transparencies pages 8, 17, 18

Notes for substitute teacher _____

_____

_____

Closing the Lesson
___ Communicating about Algebra, Student's Edition page 408
___ Guided Practice Exercises, Student's Edition page 409

Homework Assignment, pages 409–411
___ Basic/Average: Ex. 7–29 odd, 35–53 odd
___ Above Average: Ex. 11–33 odd, 37–38, 41–55 odd
___ Advanced: Ex. 13–33 odd, 37–38, 43–57 odd

Reteaching the Lesson
___ Extra Practice Copymasters page 50
___ Reteaching Copymasters page 50

Extending the Lesson
___ Technology, Teacher's Edition page 410
___ Enrichment, Teacher's Edition page 411
___ Applications Handbook page 21

Notes

_____

_____

_____

Teacher's Name _____ Class _____ Date _____ Room _____

Goals
    1. Use properties of logarithms.
    2. Use logarithms to solve real-life problems.

State/Local Objectives _____

_____

NCTM Curriculum Standards: Problem Solving, Communication, Connections, Structure

✓ **Check items you wish to use for this lesson.**

Introducing the Lesson
___ Teaching Tools: Problem of the Day copymasters page 47, or Teacher's Edition page 413
___ Teaching Tools: Warm-Up Exercises copymasters page 84, or Teacher's Edition page 413

    Notes for substitute teacher _____

_____

_____

Closing the Lesson
___ Communicating about Algebra, Student's Edition page 415
___ Guided Practice Exercises, Student's Edition page 416

Homework Assignment, pages 416–418
___ Basic/Average: Ex. 5–37 odd
___ Above Average: Ex. 5, 7, 15–37 odd
___ Advanced: Ex. 5, 7, 15–33 odd

Reteaching the Lesson
___ Extra Practice Copymasters page 51
___ Reteaching Copymasters page 51

Extending the Lesson
___ Enrichment, Teacher's Edition pages 418–419
___ Technology, Teacher's Edition page 418
___ Applications Handbook pages 34, 35, 37
___ Cultural Diversity Extensions page 34
___ Research Project: Alternative Assessment page 22

Notes

_____

_____

_____

2 days for Basic Course, 1 day for Full Course

Teacher's Name_____ Class_____ Date_____ Room_____

Goals
    1. Use the number e as the base of an exponential function.
    2. Use the natural base e in real-life problems.

State/Local Objectives _____

_____

NCTM Curriculum Standards: Problem Solving, Communication, Connections, Reasoning, Functions,
                      Calculus concepts, Structure

✓ **Check items you wish to use for this lesson.**

Introducing the Lesson
____ Teaching Tools: Problem of the Day copymasters page 48, or Teacher's Edition page 420
____ Teaching Tools: Warm-Up Exercises copymasters page 84, or Teacher's Edition page 420

Teaching the Lesson using the following:
____ Teaching Tools: transparencies pages 7, 8, 19

    Notes for substitute teacher _____

_____

_____

Closing the Lesson
____ Communicating about Algebra, Student's Edition page 422
____ Extend Communicating, Teacher's Edition page 423
____ Guided Practice Exercises, Student's Edition page 423

Homework Assignment, pages 423–425
____ Basic/Average: Ex. 9–19 odd, 25, 27, 29–32, 33, 41, 45–51 odd, 52, 58–60
____ Above Average: Ex. 9–21 odd, 25–32, 39–41, 45–52, 58–60
____ Advanced: Ex. 9–21 odd, 25–32, 37–41 odd, 45–52, 58–62

Reteaching the Lesson
____ Extra Practice Copymasters page 52
____ Reteaching Copymasters page 52

Extending the Lesson
____ Technology, Teacher's Edition page 423
____ Technology, Teacher's Edition page 424
____ Enrichment, Teacher's Edition page 425
____ Applications Handbook page 21

Notes

_____

_____

_____

Teacher's Name _____ *Class* _____ *Date* _____ *Room* _____

Goals
    1. Evaluate natural logarithmic expressions and graph natural logarithmic functions.
    2. Use natural logarithmic functions as real-life models.

State/Local Objectives _____

_____

NCTM Curriculum Standards: Problem Solving, Communication, Reasoning, Connections, Functions,
                       Calculus concepts, Geometry (algebraic)

✓ **Check items you wish to use for this lesson.**

Introducing the Lesson
___ Teaching Tools: Problem of the Day copymasters page 48, or Teacher's Edition page 426
___ Teaching Tools: Warm-Up Exercises copymasters page 84, or Teacher's Edition page 426

Teaching the Lesson using the following:
___ Color Transparencies page 44
___ Teaching Tools: transparencies pages 2, 3, 7, 8, 19

    Notes for substitute teacher _____

_____

_____

Closing the Lesson
___ Communicating about Algebra, Student's Edition page 428
___ Extend Communicating: Cooperative Learning, Teacher's Edition page 429
___ Guided Practice Exercises, Student's Edition page 429

Homework Assignment, pages 429–431
___ Basic/Average: Ex. 11–19 odd, 23, 29–32, 37–43, 45–49 odd, 51–53, 55–60, 61–67 odd
___ Above Average: Ex. 11–21 odd, 27–32, 37–43 odd, 45–49 odd, 51–56, 59–69
___ Advanced: Ex. 11–21 odd, 27–32, 37–43 odd, 45–49 odd, 51–56, 59–69 odd, 70

Reteaching the Lesson
___ Extra Practice Copymasters page 53
___ Reteaching Copymasters page 53

Extending the Lesson
___ Technology, Teacher's Edition page 430
___ Enrichment, Teacher's Edition page 431
___ Technology: Using Calculators and Computers page 50

Notes

_____

_____

_____

Teacher's Name _____ Class _____ Date _____ Room _____

Goals
    1. Solve exponential and logarithmic equations.
    2. Use exponential and logarithmic equations to answer questions about real life.

State/Local Objectives _____

_____

NCTM Curriculum Standards: Problem Solving, Communication, Connections, Geometry (algebraic),
                Structure

✓ **Check items you wish to use for this lesson.**

Introducing the Lesson
____ Teaching Tools: Problem of the Day copymasters page 48, or Teacher's Edition page 434
____ Teaching Tools: Warm-Up Exercises copymasters page 85, or Teacher's Edition page 434

Teaching the Lesson using the following:
____ Color Transparencies page 45

    Notes for substitute teacher _____

_____

_____

Closing the Lesson
____ Communicating about Algebra, Student's Edition page 437
____ Guided Practice Exercises, Student's Edition page 438

Homework Assignment, pages 438–440
____ Basic/Average: Ex. 9–21 odd, 37–45 odd, 47–59 odd, 60–63
____ Above Average: Ex. 9–23 odd, 35–45 odd, 47–51 odd, 52–54, 60–63
____ Advanced: Ex. 9–23 odd, 35–45 odd, 47–51 odd, 52–54, 60–63

Reteaching the Lesson
____ Extra Practice Copymasters page 54
____ Reteaching Copymasters page 54

Extending the Lesson
____ Technology, Teacher's Edition page 439
____ Enrichment, Teacher's Edition page 440
____ Applications Handbook page 21
____ Cooperative Learning, Alternative Assessment pages 22–23
____ Technology: Using Calculators and Computers page 52

Notes

_____

_____

_____

*Teacher's Name*_____ *Class*_____ *Date*_____ *Room*_____

Goals   1. Graph a logistics growth function.
          2. Use logistics growth functions to answer questions about real-life situations.

State/Local Objectives _____

_____

NCTM Curriculum Standards: Problem Solving, Communication, Connections, Functions

✓ **Check items you wish to use for this lesson.**

Introducing the Lesson
____ Teaching Tools: Problem of the Day copymasters page 49, or Teacher's Edition page 441
____ Teaching Tools: Warm-Up Exercises copymasters page 85, or Teacher's Edition page 441

Teaching the Lesson using the following:
____ Color Transparencies page 45
____ Teaching Tools: transparencies pages 7–8

Notes for substitute teacher _____

_____

_____

Closing the Lesson
____ Communicating about Algebra, Student's Edition page 443
____ Extend Communicating: Cooperative Learning, Teacher's Edition page 443
____ Guided Practice Exercises, Student's Edition page 444

Homework Assignment, pages 444–446
____ Basic/Average: Ex. 5–14, 23, 25, 27–32, 33–38
____ Above Average: Ex. 5–14, 21–27 odd, 28–39
____ Advanced: Ex. 5–14, 21–27 odd, 28–40

Reteaching the Lesson
____ Extra Practice Copymasters page 55
____ Reteaching Copymasters page 55

Extending the Lesson
____ Technology, Teacher's Edition page 444
____ Writing, Teacher's Edition page 444
____ Enrichment, Teacher's Edition page 446

Notes

_____

_____

_____

1 day for Basic Course, 1 day for Full Course

*Teacher's Name*_____ *Class*_____ *Date*_____ *Room*_____

Goals
1. Add, subtract, and multiply polynomials.
2. Use polynomial operations to solve real-life problems.

State/Local Objectives _____

_____

NCTM Curriculum Standards: Problem Solving, Communication, Connections

✓ **Check items you wish to use for this lesson.**

Introducing the Lesson
____ Teaching Tools: Problem of the Day copymasters page 49, or Teacher's Edition page 456
____ Teaching Tools: Warm-Up Exercises copymasters page 85, or Teacher's Edition page 456

Teaching the Lesson using the following:
____ Color Transparencies page 47
____ Teaching Tools: transparencies page 22

Notes for substitute teacher _____

_____

_____

Closing the Lesson
____ Communicating about Algebra, Student's Edition page 459
____ Extend Communicating, Teacher's Edition page 459
____ Guided Practice Exercises, Student's Edition page 460

Homework Assignment, pages 460–462
____ Basic/Average: Ex. 9–29 odd, 41–53 odd, 60–62, 65–71, 78
____ Above Average: Ex. 9–31 odd, 41–53 odd, 59–62, 69–71, 76–78
____ Advanced: Ex. 9–31 odd, 41–53 odd, 59–64, 69–71, 77–80

Reteaching the Lesson
____ Extra Practice Copymasters page 56
____ Reteaching Copymasters page 56
____ Alternative Assessment: Math Log pages 54–56

Extending the Lesson
____ Enrichment, Teacher's Edition page 462
____ Alternative Assessment page 23

Notes

_____

_____

_____

Lesson Plan 9.2
pages 463–469

Teacher's Name _____ Class _____ Date _____ Room _____

Goals
 1. Sketch the graph of a polynomial function.
 2. Use polynomial functions as models of real-life situations.

State/Local Objectives _____
_____

NCTM Curriculum Standards: Problem Solving, Communication, Connections, Functions,
 Geometry (algebraic), Calculus concepts

✓ **Check items you wish to use for this lesson.**

Introducing the Lesson
___ Teaching Tools: Problem of the Day copymasters page 50, or Teacher's Edition page 463
___ Teaching Tools: Warm-Up Exercises copymasters page 86, or Teacher's Edition page 463

Teaching the Lesson using the following:
___ Color Transparencies page 48
___ Teaching Tools: transparencies pages 2, 3, 7, 8, 20

 Notes for substitute teacher _____
_____
_____

Closing the Lesson
___ Communicating about Algebra, Student's Edition page 466
___ Extend Communicating, Teacher's Edition page 466
___ Guided Practice Exercises, Student's Edition page 467

Homework Assignment, pages 467–469
___ Basic/Average: Ex. 7–19 odd, 25–30, 37–43, 51–55 odd, 56, 63–64
___ Above Average: Ex. 7–19 odd, 25–30, 35–45, 51–56, 63–64
___ Advanced: Ex. 7–19 odd, 25–30, 35–45, 51–56, 63–68

Reteaching the Lesson
___ Extra Practice Copymasters page 57
___ Reteaching Copymasters page 57

Extending the Lesson
___ Technology, Teacher's Edition page 468
___ Writing, Teacher's Edition page 468
___ Enrichment, Teacher's Edition page 469
___ Technology: Using Calculators and Computers page 54

Notes

_____
_____
_____

Teacher's Name_____ Class_____ Date_____ Room_____

Goals
1. Factor polynomial expressions and equations.
2. Use factoring to solve real-life problems.

State/Local Objectives _____

_____

NCTM Curriculum Standards: Problem Solving, Communication, Reasoning, Connections

✓ **Check items you wish to use for this lesson.**

Introducing the Lesson
___ Teaching Tools: Problem of the Day copymasters page 50, or Teacher's Edition page 473
___ Teaching Tools: Warm-Up Exercises copymasters page 86, or Teacher's Edition page 473

Teaching the Lesson using the following:
___ Common-Error Alert, Teacher's Edition page 474
___ Color Transparencies pages 48, 49

  Notes for substitute teacher _____

_____

_____

Closing the Lesson
___ Communicating about Algebra, Student's Edition page 476
___ Extend Communicating: Cooperative Learning, Teacher's Edition page 476
___ Guided Practice Exercises, Student's Edition page 477

Homework Assignment, pages 477–479
___ Basic/Average: Ex. 7–11 odd, 15–45 odd, 51–55 odd, 62–67, 69–77 odd
___ Above Average: Ex. 11–49 odd, 51–57 odd, 62–67, 69–81 odd
___ Advanced: Ex. 11–49 odd, 51–61 odd, 62–68, 69–75 odd, 79–83

Reteaching the Lesson
___ Extra Practice Copymasters page 58
___ Reteaching Copymasters page 58

Extending the Lesson
___ Enrichment, Teacher's Edition pages 478–479
___ Applications Handbook pages 46–48
___ Technology: Using Calculators and Computers page 57

Notes

_____

_____

_____

ⓒ D.C. Heath and Company

Teacher's Name _____ Class _____ Date _____ Room _____

Goals
1. Divide polynomials using long division and synthetic division and relate the quotient to the Remainder Theorem and the Factor Theorem.
2. Use polynomial division in real-life problems.

State/Local Objectives _____

_____

NCTM Curriculum Standards: Problem Solving, Communication, Connections

✓ **Check items you wish to use for this lesson.**

Introducing the Lesson
___ Teaching Tools: Problem of the Day copymasters page 50, or Teacher's Edition page 480
___ Teaching Tools: Warm-Up Exercises copymasters page 86, or Teacher's Edition page 480

Teaching the Lesson using the following:
___ Color Transparencies pages 49, 50, 51

   Notes for substitute teacher _____

_____

_____

Closing the Lesson
___ Communicating about Algebra, Student's Edition page 483
___ Guided Practice Exercises, Student's Edition page 484

Homework Assignment, pages 484–486
___ Basic/Average: Ex. 9–15 odd, 23–33 odd, 39, 43–49 odd, 51–57, 58, 61
___ Above Average: Ex. 9–27 odd, 33–39 odd, 43–50, 57, 58, 61
___ Advanced: Ex. 9–27 odd, 33–39 odd, 43–50, 57–61

Reteaching the Lesson
___ Extra Practice Copymasters page 59
___ Reteaching Copymasters page 59

Extending the Lesson
___ Technology, Teacher's Edition page 485
___ Enrichment, Teacher's Edition page 486
___ Problem Solving, Alternative Assessment page 23
___ Cooperative Learning, Alternative Assessment page 23
___ Technology: Using Calculators and Computers page 59

Notes

_____

_____

_____

*Teacher's Name*_____ *Class*_____ *Date*_____ *Room*_____

Goals   1. Find the rational zeros of a polynomial function.
2. Use the zeros of a polynomial function to solve real-life problems.

State/Local Objectives _____

_____

NCTM Curriculum Standards: Problem Solving, Communication, Reasoning, Connections, Functions

✓ **Check items you wish to use for this lesson.**

Introducing the Lesson
___ Teaching Tools: Problem of the Day copymasters page 51, or Teacher's Edition page 488
___ Teaching Tools: Warm-Up Exercises copymasters page 87, or Teacher's Edition page 488

Teaching the Lesson using the following:
___ Color Transparencies pages 51, 52
___ Teaching Tools: transparencies page 8

    Notes for substitute teacher _____

_____

_____

Closing the Lesson
___ Communicating about Algebra, Student's Edition page 490
___ Extend Communicating: Cooperative Learning, Teacher's Edition page 490
___ Guided Practice Exercises, Student's Edition page 491

Homework Assignment, pages 491–493
___ Basic/Average: Ex. 9–17 odd, 27–31 odd, 32, 34, 43–47 odd, 51–52
___ Above Average: Ex. 9–15 odd, 23–31 odd, 32–35, 45–51 odd, 52, 53
___ Advanced: Ex. 9–15 odd, 23–31 odd, 32–34, 49–53, 55, 56

Reteaching the Lesson
___ Extra Practice Copymasters page 60
___ Reteaching Copymasters page 60

Extending the Lesson
___ Technology, Teacher's Edition page 491
___ Enrichment, Teacher's Edition page 493
___ Technology: Using Calculators and Computers page 61

Notes

_____

_____

_____

Teacher's Name _____ Class _____ Date _____ Room _____

Goals   1. Look at zeros of polynomial functions, factors of polynomials, and solutions of polynomial equations to see their connections.
2. Use polynomials to solve real-life problems.

State/Local Objectives _____

_____

NCTM Curriculum Standards: Problem Solving, Communication, Reasoning, Connections, Functions, Structure

✓ **Check items you wish to use for this lesson.**

Introducing the Lesson
___ Teaching Tools: Problem of the Day copymasters page 51, or Teacher's Edition page 494
___ Teaching Tools: Warm-Up Exercises copymasters page 87, or Teacher's Edition page 494

Notes for substitute teacher _____

_____

Closing the Lesson
___ Communicating about Algebra, Student's Edition page 497
___ Extend Communicating: Cooperative Learning, Teacher's Edition page 497
___ Guided Practice Exercises, Student's Edition page 498

Homework Assignment, pages 498–500
___ Basic/Average: Ex. 7–15 odd, 25–35 odd, 36–43, 51–55
___ Above Average: Ex. 7–21 odd, 25–35 odd, 36–47, 54–56, 58
___ Advanced: Ex. 7–21 odd, 25–35 odd, 36–46, 54–58

Reteaching the Lesson
___ Extra Practice Copymasters page 61
___ Reteaching Copymasters page 61

Extending the Lesson
___ Enrichment, Teacher's Edition page 500
___ Alternative Assessment page 24
___ Technology: Using Calculators and Computers page 65

Notes

_____

_____

_____

Teacher's Name _____ *Class* _____ *Date* _____ *Room* _____

Goals
1. Find the range of a collection of numbers.
2. Find the standard deviation of a collection of numbers.

State/Local Objectives _____
_____

NCTM Curriculum Standards: Problem Solving, Communication, Connections, Statistics, Discrete Math

✓ **Check items you wish to use for this lesson.**

Introducing the Lesson
____ Teaching Tools: Problem of the Day copymasters page 51, or Teacher's Edition page 501
____ Teaching Tools: Warm-Up Exercises copymasters page 87, or Teacher's Edition page 501

Teaching the Lesson using the following:
____ Color Transparencies pages 52, 53
____ Teaching Tools: transparencies page 5

Notes for substitute teacher _____
_____
_____

Closing the Lesson
____ Communicating about Algebra, Student's Edition page 503
____ Extend Communicating: Cooperative Learning, Teacher's Edition page 503
____ Guided Practice Exercises, Student's Edition page 504

Homework Assignment, pages 504–507
____ Basic/Average: Ex. 5–9 odd, 11–17, 21–26
____ Above Average: Ex. 5–9 odd, 10–21, 24–26
____ Advanced: Ex. 5–9 odd, 10–21, 24–25

Reteaching the Lesson
____ Extra Practice Copymasters page 62
____ Reteaching Copymasters page 62

Extending the Lesson
____ Technology, Teacher's Edition page 502
____ Technology, Teacher's Edition page 504
____ Enrichment, Teacher's Edition page 507
____ Applications Handbook page 11
____ Cultural Diversity Extensions page 35
____ Technology: Using Calculators and Computers page 68

Notes
_____
_____
_____

Teacher's Name _____ *Class* _____ *Date* _____ *Room* _____

Goals   1. Graph a rational function using asymptotes.
            2. Use rational functions as real-life models.

State/Local Objectives _____

_____

NCTM Curriculum Standards: Problem Solving, Communication, Connections, Functions

✓ **Check items you wish to use for this lesson.**

Introducing the Lesson
____ Teaching Tools: Problem of the Day copymasters page 52, or Teacher's Edition page 518
____ Teaching Tools: Warm-Up Exercises copymasters page 88, or Teacher's Edition page 518

Teaching the Lesson using the following:
____ Color Transparencies page 54
____ Teaching Tools: transparencies pages 2, 3, 7, 8, 21

       Notes for substitute teacher _____

_____

_____

Closing the Lesson
____ Communicating about Algebra, Student's Edition page 520
____ Guided Practice Exercises, Student's Edition page 521

Homework Assignment, pages 521–523
____ Basic/Average: Ex. 9–21 odd, 27–30, 31–37 odd, 43–45, 53–55
____ Above Average: Ex. 9–23 odd, 26–30, 31–37 odd, 43–46, 53–55
____ Advanced: Ex. 9–23 odd, 26–30, 31–37 odd, 43–46, 53–57

Reteaching the Lesson
____ Extra Practice Copymasters page 63
____ Reteaching Copymasters page 63
____ Alternative Assessment: Math Log pages 57–59

Extending the Lesson
____ Technology, Teacher's Edition page 522
____ Writing, Teacher's Edition page 522
____ Enrichment, Teacher's Edition page 523
____ Research Project: Alternative Assessment page 24

Notes

_____

_____

_____

Teacher's Name _____ Class _____ Date _____ Room _____

Goals
1. Create and use real-life models using inverse variation.
2. Create and use models of real-life situations that use joint variation.

State/Local Objectives _____

_____

NCTM Curriculum Standards: Problem Solving, Communication, Connections, Calculus concepts

✓ **Check items you wish to use for this lesson.**

Introducing the Lesson
____ Teaching Tools: Problem of the Day copymasters page 52, or Teacher's Edition page 526
____ Teaching Tools: Warm-Up Exercises copymasters page 88, or Teacher's Edition page 526

Teaching the Lesson using the following:
____ Color Transparencies page 55
____ Extension for Example 3: Research, Teacher's Edition page 527

Notes for substitute teacher _____

_____

_____

Closing the Lesson
____ Communicating about Algebra, Student's Edition page 528
____ Extend Communicating, Teacher's Edition page 528
____ Guided Practice Exercises, Student's Edition page 529

Homework Assignment, pages 529–531
____ Basic/Average: Ex. 7–15 odd, 17, 18, 22–24, 30–35
____ Above Average: Ex. 7–19 odd, 20–24, 30–35
____ Advanced: Ex. 7–19 odd, 20–24, 30–35

Reteaching the Lesson
____ Extra Practice Copymasters page 64
____ Reteaching Copymasters page 64

Extending the Lesson
____ Enrichment, Teacher's Edition page 531
____ Applications Handbook page 57
____ Cultural Diversity Extensions page 36

Notes

_____

_____

Teacher's Name _____ *Class* _____ *Date* _____ *Room* _____

Goals
1. Multiply and divide rational expressions, writing the result in simplest form.
2. Use rational expressions as real-life models.

State/Local Objectives _____

_____

NCTM Curriculum Standards: Problem Solving, Communication, Reasoning, Connections, Geometry (synthetic)

✓ **Check items you wish to use for this lesson.**

Introducing the Lesson
___ Teaching Tools: Problem of the Day copymasters page 52, or Teacher's Edition page 533
___ Teaching Tools: Warm-Up Exercises copymasters page 89, or Teacher's Edition page 533

Teaching the Lesson using the following:
___ Extension for Example 7, Teacher's Edition page 534
___ Common-Error Alert, Teacher's Edition page 535

Notes for substitute teacher _____

_____

_____

Closing the Lesson
___ Communicating about Algebra, Student's Edition page 536
___ Guided Practice Exercises, Student's Edition page 537

Homework Assignment, pages 537–539
___ Basic/Average: Ex. 7–15 odd, 25–33 odd, 42–46, 57–59
___ Above Average: Ex. 7–17 odd, 27–35 odd, 42–46, 50, 57–59
___ Advanced: Ex. 9–19 odd, 27–37 odd, 42–46, 52, 57–60

Reteaching the Lesson
___ Extra Practice Copymasters page 65
___ Reteaching Copymasters page 65

Extending the Lesson
___ Enrichment, Teacher's Edition page 539
___ Cooperative Learning, Alternative Assessment page 24

Notes

_____

_____

_____

2 days for Basic Course, 1 day for Full Course

Teacher's Name_____ Class_____ Date_____ Room_____

Goals
1. Solve equations that contain rational expressions.
2. Use rational equations to solve real-life problems.

State/Local Objectives _____

_____

NCTM Curriculum Standards: Problem Solving, Communication, Connections

✓ **Check items you wish to use for this lesson.**

Introducing the Lesson
____ Teaching Tools: Problem of the Day copymasters page 53, or Teacher's Edition page 541
____ Teaching Tools: Warm-Up Exercises copymasters page 89, or Teacher's Edition page 541

Teaching the Lesson using the following:
____ Common-Error Alert, Teacher's Edition page 542
____ Extension for Examples 3–4, Teacher's Edition page 542
____ Technology for Examples 6–7, Teacher's Edition page 542
____ Color Transparencies page 56
____ Teaching Tools: transparencies page 8

Notes for substitute teacher _____

_____

_____

Closing the Lesson
____ Communicating about Algebra, Student's Edition page 544
____ Guided Practice Exercises, Student's Edition page 545

Homework Assignment, pages 545–547
____ Basic/Average: Ex. 7, 11–23 odd, 31–41 odd, 49, 51–60
____ Above Average: Ex. 7–23 odd, 33–43 odd, 49–60
____ Advanced: Ex. 7–23 odd, 33–43 odd, 49–61

Reteaching the Lesson
____ Extra Practice Copymasters page 66
____ Reteaching Copymasters page 66

Extending the Lesson
____ Technology, Teacher's Edition page 546
____ Enrichment, Teacher's Edition page 547
____ Technology: Using Calculators and Computers page 71

Notes

_____

_____

_____

Teacher's Name _____ *Class* _____ *Date* _____ *Room* _____

Goals   1. Add and subtract rational expressions and simplify complex fractions.
2. Use rational expressions as models of real-life situations.

State/Local Objectives _____

_____

NCTM Curriculum Standards: Problem Solving, Communication, Connections

✓ **Check items you wish to use for this lesson.**

Introducing the Lesson
___ Teaching Tools: Problem of the Day copymasters page 53, or Teacher's Edition page 548
___ Teaching Tools: Warm-Up Exercises copymasters page 89, or Teacher's Edition page 548

Teaching the Lesson using the following:
___ Technology for Example 4, Teacher's Edition page 549

Notes for substitute teacher _____

_____

_____

Closing the Lesson
___ Communicating about Algebra, Student's Edition page 550
___ Guided Practice Exercises, Student's Edition page 551

Homework Assignment, pages 551–554
___ Basic/Average: Ex. 5, 9–17 odd, 29–35 odd, 41–44, 47–57 odd, 59
___ Above Average: Ex. 5–17 odd, 27–37 odd, 41–45, 56–59
___ Advanced: Ex. 7–19 odd, 27–37 odd, 41–45, 56–59

Reteaching the Lesson
___ Extra Practice Copymasters page 67
___ Reteaching Copymasters page 67

Extending the Lesson
___ Enrichment, Teacher's Edition page 554

Notes

_____

_____

_____

Teacher's Name _____ Class _____ Date _____ Room _____

Goals   1. Construct an amortization table for a loan.
            2. Find the monthly paymnent for an installment loan.

State/Local Objectives _____

_____

NCTM Curriculum Standards: Problem Solving, Communication, Connections, Discrete Math

✓ **Check items you wish to use for this lesson.**

Introducing the Lesson
____ Teaching Tools: Problem of the Day copymasters page 54, or Teacher's Edition page 555
____ Teaching Tools: Warm-Up Exercises copymasters page 90, or Teacher's Edition page 555

Teaching the Lesson using the following:
____ Color Transparencies page 57

   Notes for substitute teacher _____

_____

_____

Closing the Lesson
____ Communicating about Algebra, Student's Edition page 557
____ Extend Communicating: Cooperative Learning, Teacher's Edition page 557
____ Guided Practice Exercises, Student's Edition page 558

Homework Assignment, pages 558–560
____ Basic/Average: Ex. 7, 11, 15, 19, 21–24, 26, 27, 29, 31–34
____ Above Average: Ex. 5–19 odd, 21–25, 27–33 odd, 35
____ Advanced: Ex. 5–19 odd, 21–27, 29–35 odd, 36

Reteaching the Lesson
____ Extra Practice Copymasters page 68
____ Reteaching Copymasters page 68

Extending the Lesson
____ Technology, Teacher's Edition page 558
____ Enrichment, Teacher's Edition page 560
____ Technology: Using Calculators and Computers pages 73, 75

Notes

_____

_____

_____

Teacher's Name _____ Class _____ Date _____ Room _____

Goals
1. Write an equation of a parabola and sketch its graph.
2. Use parabolas to solve real-life problems.

State/Local Objectives _____

_____

NCTM Curriculum Standards: Problem Solving, Communication, Connections, Geometry (algebraic)

✓ **Check items you wish to use for this lesson.**

Introducing the Lesson
___ Teaching Tools: Problem of the Day copymasters page 54, or Teacher's Edition page 570
___ Teaching Tools: Warm-Up Exercises copymasters page 90, or Teacher's Edition page 570

Teaching the Lesson using the following:
___ Color Transparencies page 58
___ Teaching Tools: transparencies pages 7, 8, 13

Notes for substitute teacher _____

_____

_____

Closing the Lesson
___ Communicating about Algebra, Student's Edition page 572
___ Guided Practice Exercises, Student's Edition page 573

Homework Assignment, pages 573–575
___ Basic/Average: Ex. 13–23 odd, 29–37 odd, 39–42, 47–55 odd, 57, 67–69
___ Above Average: Ex. 13–23 odd, 29–37 odd, 39–42, 45–55 odd, 56–59, 67–70
___ Advanced: Ex. 13–23 odd, 29–37 odd, 39–42, 45–55 odd, 56–60, 67–72

Reteaching the Lesson
___ Extra Practice Copymasters page 69
___ Reteaching Copymasters page 69
___ Alternative Assessment: Math Log pages 60–62

Extending the Lesson
___ Technology, Teacher's Edition page 574
___ Enrichment, Teacher's Edition page 575
___ Research Project: Alternative Assessment page 25

Notes

_____

_____

_____

Teacher's Name _____ Class _____ Date _____ Room _____

Goals   1. Write an equation of a circle and sketch its graph.
2. Use equations and graphs of circles in real-life problems.

State/Local Objectives _____

_____

NCTM Curriculum Standards: Problem Solving, Communication, Reasoning, Connections, Geometry (algebraic)

✓ **Check items you wish to use for this lesson.**

Introducing the Lesson
____ Teaching Tools: Problem of the Day copymasters page 54, or Teacher's Edition page 576
____ Teaching Tools: Warm-Up Exercises copymasters page 91, or Teacher's Edition page 576

Teaching the Lesson using the following:
____ Color Transparencies pages 58, 59
____ Teaching Tools: transparencies pages 7, 8

Notes for substitute teacher _____

_____

_____

Closing the Lesson
____ Communicating about Algebra, Student's Edition page 578
____ Extend Communicating, Teacher's Edition page 578
____ Guided Practice Exercises, Student's Edition page 579

Homework Assignment, pages 579–581
____ Basic/Average: Ex. 7, 11, 13–16, 21–31 odd, 38, 42, 53, 55–60
____ Above Average: Ex. 7, 9, 13–16, 19–33 odd, 37–41, 53–61
____ Advanced: Ex. 7, 9, 13–16, 19–33 odd, 37–42, 54–61

Reteaching the Lesson
____ Extra Practice Copymasters page 70
____ Reteaching Copymasters page 70

Extending the Lesson
____ Enrichment, Teacher's Edition page 581
____ Applications Handbook page 57
____ Technology: Using Calculators and Computers page 77

Notes

_____

_____

_____

Teacher's Name _____ *Class* _____ *Date* _____ *Room* _____

Goals   1. Write an equation of an ellipse and sketch its graph.
             2. Use ellipses to solve real-life problems.

State/Local Objectives _____

_____

NCTM Curriculum Standards: Problem Solving, Communication, Connections, Geometry (algebraic)

✓ **Check items you wish to use for this lesson.**

Introducing the Lesson
___ Teaching Tools: Problem of the Day copymasters page 55, or Teacher's Edition page 583
___ Teaching Tools: Warm-Up Exercises copymasters page 91, or Teacher's Edition page 583

Teaching the Lesson using the following:
___ Teaching Tools: transparencies pages 7, 8

  Notes for substitute teacher _____

_____

_____

Closing the Lesson
___ Communicating about Algebra, Student's Edition page 585
___ Guided Practice Exercises, Student's Edition page 586

Homework Assignment, pages 586–588
___ Basic/Average: Ex. 11, 17, 21, 25–28, 31–39 odd, 47–51, 56, 63–67
___ Above Average: Ex. 11, 17, 21, 25–28, 31–39 odd, 45–54, 56, 66–68
___ Advanced: Ex. 15–25 odd, 26–29, 31–43 odd, 47–56, 68

Reteaching the Lesson
___ Extra Practice Copymasters page 71
___ Reteaching Copymasters page 71

Extending the Lesson
___ Research, Teacher's Edition page 587
___ Enrichment, Teacher's Edition page 588
___ Technology: Using Calculators and Computers page 79

Notes

_____

_____

_____

2 days for Basic Course, 1 day for Full Course

Teacher's Name_____ Class_____ Date_____ Room_____

Goals
1. Write the equation of a hyperbola and graph it, using asymptotes.
2. Use hyperbolas to solve real-life problems.

State/Local Objectives _____

_____

NCTM Curriculum Standards: Problem Solving, Communication, Connections, Geometry (algebraic)

✓ **Check items you wish to use for this lesson.**

Introducing the Lesson
___ Teaching Tools: Problem of the Day copymasters page 55, or Teacher's Edition page 590
___ Teaching Tools: Warm-Up Exercises copymasters page 91, or Teacher's Edition page 590

Teaching the Lesson using the following:
___ Color Transparencies page 59
___ Teaching Tools: transparencies pages 7, 8

Notes for substitute teacher _____

_____

_____

Closing the Lesson
___ Communicating about Algebra, Student's Edition page 593
___ Guided Practice Exercises, Student's Edition page 594

Homework Assignment, pages 594–596
___ Basic/Average: Ex. 11–19 odd, 25–28, 33, 37, 43, 49, 52–62
___ Above Average: Ex. 9–19 odd, 24–28, 31–43 odd, 47, 44–59, 51–52, 57–63
___ Advanced: Ex. 9–19 odd, 24–28, 31–43 odd, 47, 49–53, 60–64

Reteaching the Lesson
___ Extra Practice Copymasters page 72
___ Reteaching Copymasters page 72

Extending the Lesson
___ Enrichment, Teacher's Edition page 596
___ Technology, Teacher's Edition page 596

Notes

_____

_____

_____

Teacher's Name _____ Class _____ Date _____ Room _____

Goals
  1. Write an equation of a parabola with a vertex at (h, k) and an equation of a circle, an ellipse, or a hyperbola with its center at (h, k).
  2. Use translations to solve real-life problems.

State/Local Objectives _____

_____

NCTM Curriculum Standards: Problem Solving, Communication, Connections, Geometry (algebraic perspective)

✓ **Check items you wish to use for this lesson.**

Introducing the Lesson
___ Teaching Tools: Problem of the Day copymasters page 55, or Teacher's Edition page 597
___ Teaching Tools: Warm-Up Exercises copymasters page 92, or Teacher's Edition page 597

Teaching the Lesson using the following:
___ Color Transparencies page 60
___ Teaching Tools: transparencies pages 2, 3, 7, 8

   Notes for substitute teacher _____

_____

_____

Closing the Lesson
___ Communicating about Algebra, Student's Edition page 600
___ Guided Practice Exercises, Student's Edition page 601

Homework Assignment, pages 601–603
___ Basic/Average: Ex. 7–14, 15–23 odd, 27, 31, 35, 39–42, 43–51 odd
___ Above Average: Ex. 7–14, 15–25 odd, 29, 33, 37, 39–42, 52
___ Advanced: Ex. 7–14, 15–25 odd, 29, 33, 37, 39–42, 50–52

Reteaching the Lesson
___ Extra Practice Copymasters page 73
___ Reteaching Copymasters page 73

Extending the Lesson
___ Writing, Teacher's Edition page 601
___ Enrichment, Teacher's Edition page 602
___ Cultural Diversity Extensions page 37
___ Technology: Using Calculators and Computers page 81

Notes

_____

_____

_____

Teacher's Name_____ Class_____ Date_____ Room_____

Goals
1. Classify a conic from its general equation.
2. Use equations of conics to solve mathematical and real-life problems.

State/Local Objectives _____

_____

NCTM Curriculum Standards: Problem Solving, Communication, Reasoning, Connections, Geometry (algebraic)

✓ **Check items you wish to use for this lesson.**

Introducing the Lesson
___ Teaching Tools: Problem of the Day copymasters page 56, or Teacher's Edition page 606
___ Teaching Tools: Warm-Up Exercises copymasters page 92, or Teacher's Edition page 606

Teaching the Lesson using the following:
___ Color Transparencies pages 60, 61
___ Teaching Tools: transparencies pages 7, 8, 23

  Notes for substitute teacher _____

_____

_____

Closing the Lesson
___ Communicating about Algebra, Student's Edition page 609
___ Extend Communicating, Teacher's Edition page 609
___ Guided Practice Exercises, Student's Edition page 610

Homework Assignment, pages 610–613
___ Basic/Average: Ex. 7–21 odd, 23–30, 31, 37, 41, 43–53 odd, 55, 59, 61–64
___ Above Average: Ex. 7–21 odd, 23–30, 31–39 odd, 43–53 odd, 55–66
___ Advanced: Ex. 7–21 odd, 23–30, 31–39 odd, 43–53 odd, 55–68

Reteaching the Lesson
___ Extra Practice Copymasters page 74
___ Reteaching Copymasters page 74

Extending the Lesson
___ Writing, Teacher's Edition page 611
___ Enrichment, Teacher's Edition page 613
___ Cooperative Learning, Alternative Assessment page 25
___ Technology: Using Calculators and Computers pages 34, 35

Notes

_____

_____

_____

Teacher's Name _____ Class _____ Date _____ Room _____

Goals
1. Write and use sequences and series.
2. Use sequences and series as models of real-life situations.

State/Local Objectives _____

_____

NCTM Curriculum Standards: Problem Solving, Communication, Reasoning, Connections, Calculus concepts, Structure

✓ **Check items you wish to use for this lesson.**

Introducing the Lesson
___ Teaching Tools: Problem of the Day copymasters page 56, or Teacher's Edition page 622
___ Teaching Tools: Warm-Up Exercises copymasters page 92, or Teacher's Edition page 622

Teaching the Lesson using the following:
___ Color Transparencies pages 63

Notes for substitute teacher _____

_____

_____

Closing the Lesson
___ Communicating about Algebra, Student's Edition page 625
___ Guided Practice Exercises, Student's Edition page 626

Homework Assignment, pages 626–628
___ Basic/Average: Ex. 9–21 odd, 29, 35–41 odd, 52, 55–58, 63
___ Above Average: Ex. 9–23 odd, 29, 35–43 odd, 47–50, 52, 54, 61–67 odd
___ Advanced: Ex. 9–23 odd, 35–43 odd, 47–50, 52, 54, 60–73, 66–68

Reteaching the Lesson
___ Extra Practice Copymasters page 75
___ Reteaching Copymasters page 75
___ Alternative Assessment: Math Log pages 63–65

Extending the Lesson
___ Enrichment, Teacher's Edition page 628
___ Cultural Diversity Extensions page 38
___ Research Project: Alternative Assessment page 26

Notes

_____

_____

_____

Teacher's Name_____ Class_____ Date_____ Room_____

Goals   1. Find the nth term and the sum of an arithmetic series.
          2. Use arithmetic series in real-life problems.

State/Local Objectives _____

_____

NCTM Curriculum Standards: Problem Solving, Communication, Connections, Calculus concepts, Structure

✓ **Check items you wish to use for this lesson.**

Introducing the Lesson
____ Teaching Tools: Problem of the Day copymasters page 57, or Teacher's Edition page 629
____ Teaching Tools: Warm-Up Exercises copymasters page 93, or Teacher's Edition page 629

Teaching the Lesson using the following:
____ Color Transparencies pages 63

    Notes for substitute teacher _____

_____

_____

Closing the Lesson
____ Communicating about Algebra, Student's Edition page 632
____ Extend Communicating: Cooperative Learning, Teacher's Edition page 632
____ Guided Practice Exercises, Student's Edition page 633

Homework Assignment, pages 633–635
____ Basic/Average: Ex. 9–27 odd, 28–31, 35–47 odd, 51–61 odd, 62
____ Above Average: Ex. 9–27 odd, 28–31, 35–45 odd, 46–48, 55–62
____ Advanced: Ex. 9–27 odd, 28–31, 35–43 odd, 44–48, 57–62

Reteaching the Lesson
____ Extra Practice Copymasters page 76
____ Reteaching Copymasters page 76

Extending the Lesson
____ Writing, Teacher's Edition page 633
____ Enrichment, Teacher's Edition page 635

Notes

_____

_____

_____

Teacher's Name _____ Class _____ Date _____ Room _____

Goals　　　1. Find the nth term and the sum of a geometric series.
　　　　　　2. Use geometric series in real-life problems.

State/Local Objectives _____

_____

NCTM Curriculum Standards: Problem Solving, Communication, Connections, Calculus concepts, Structure

✓ **Check items you wish to use for this lesson.**

Introducing the Lesson
____ Teaching Tools: Problem of the Day copymasters page 57, or Teacher's Edition page 637
____ Teaching Tools: Warm-Up Exercises copymasters page 93, or Teacher's Edition page 637

Teaching the Lesson using the following:
____ Common-Error Alert, Teacher's Edition page 638
____ Color Transparencies pages 64, 65

　　　Notes for substitute teacher _____

_____

_____

Closing the Lesson
____ Communicating about Algebra, Student's Edition page 640
____ Extend Communicating, Teacher's Edition page 640
____ Guided Practice Exercises, Student's Edition page 641

Homework Assignment, pages 641–643
____ Basic/Average: Ex. 9–13 odd, 21–29 odd, 33–41 odd, 47–53 odd, 65–71
____ Above Average: Ex. 9–17 odd, 21–27 odd, 21–43 odd, 47–53 odd, 54, 68–71
____ Advanced: Ex. 9–17 odd, 23–43 odd, 45–51 odd, 52–54, 68–71

Reteaching the Lesson
____ Extra Practice Copymasters page 77
____ Reteaching Copymasters page 77

Extending the Lesson
____ Enrichment, Teacher's Edition page 643
____ Cooperative Learning, Alternative Assessment page 27

Notes

_____

_____

_____

Teacher's Name_____ Class_____ Date_____ Room_____

Goals
1. Find the sum of an infinite geometric series.
2. Use infinite geometric series as models of real-life problems.

State/Local Objectives _____

_____

NCTM Curriculum Standards: Problem Solving, Communication, Connections, Calculus concepts, Structure

✓ **Check items you wish to use for this lesson.**

Introducing the Lesson
___ Teaching Tools: Problem of the Day copymasters page 57, or Teacher's Edition page 645
___ Teaching Tools: Warm-Up Exercises copymasters page 94, or Teacher's Edition page 645

Notes for substitute teacher _____

_____

_____

Closing the Lesson
___ Communicating about Algebra, Student's Edition page 647
___ Extend Communicating:Writing Problems, Teacher's Edition page 648
___ Guided Practice Exercises, Student's Edition page 648

Homework Assignment, pages 648–650
___ Basic/Average: Ex. 5–23 odd, 29, 30, 33, 34, 35–45 odd, 47
___ Above Average: Ex. 5–29 odd, 30–34, 35–51 odd
___ Advanced: Ex. 5–29 odd, 30–35, 42–47, 50, 51

Reteaching the Lesson
___ Extra Practice Copymasters page 78
___ Reteaching Copymasters page 78

Extending the Lesson
___ Writing, Teacher's Edition page 648
___ Enrichment, Teacher's Edition page 650
___ Technology: Using Calculators and Computers page 83

Notes

_____

_____

_____

Teacher's Name _____ *Class* _____ *Date* _____ *Room* _____

Goals   1. Use the Binomial Theorem to expand a binomial that is raised to a power.
              2. Use the Binomial Theorem in real-life situations.

State/Local Objectives _____

_____

NCTM Curriculum Standards: Problem Solving, Communication, Reasoning, Connections, Structure

✓ **Check items you wish to use for this lesson.**

Introducing the Lesson
____ Teaching Tools: Problem of the Day copymasters page 58, or Teacher's Edition page 651
____ Teaching Tools: Warm-Up Exercises copymasters page 94, or Teacher's Edition page 651

Teaching the Lesson using the following:
____ Color Transparencies page 65
____ Teaching Tools: transparencies page 24

      Notes for substitute teacher _____

_____

_____

Closing the Lesson
____ Communicating about Algebra, Student's Edition page 654
____ Guided Practice Exercises, Student's Edition page 655

Homework Assignment, pages 655–657
____ Basic/Average: Ex. 9–33 odd, 40–42, 47, 48, 51, 52
____ Above Average: Ex. 9–33 odd, 37–42, 47–52
____ Advanced: Ex. 9–35 odd, 37–42, 47–52

Reteaching the Lesson
____ Extra Practice Copymasters page 79
____ Reteaching Copymasters page 79

Extending the Lesson
____ Technology, Teacher's Edition page 655
____ Writing, Teacher's Edition page 655
____ Enrichment, Teacher's Edition page 657
____ Technology: Using Calculators and Computers page 87

Notes

_____

_____

_____

Teacher's Name _____ Class _____ Date _____ Room _____

Goals
1. Find the balance of an increasing annuity.
2. Find the monthly deposit to reach a specified balance in an annuity.

State/Local Objectives _____

_____

NCTM Curriculum Standards: Problem Solving, Communication, Connections, Discrete Math

✓ **Check items you wish to use for this lesson.**

Introducing the Lesson
___ Teaching Tools: Problem of the Day copymasters page 58, or Teacher's Edition page 658
___ Teaching Tools: Warm-Up Exercises copymasters page 94, or Teacher's Edition page 658

Teaching the Lesson using the following:
___ Color Transparencies page 66
___ Teaching Tools: transparencies pages 2, 3

___ Notes for substitute teacher _____

_____

_____

Closing the Lesson
___ Communicating about Algebra, Student's Edition page 660
___ Guided Practice Exercises, Student's Edition page 661

Homework Assignment, pages 661–663
___ Basic/Average: Ex. 9, 13–21 odd, 25–27
___ Above Average: Ex. 9, 13, 15–20, 24–28
___ Advanced: Ex. 7, 11, 15–21, 24–28

Reteaching the Lesson
___ Extra Practice Copymasters page 80
___ Reteaching Copymasters page 80

Extending the Lesson
___ Enrichment, Teacher's Edition page 663
___ Technology, Teacher's Edition page 663
___ Technology: Using Calculators and Computers page 89

Notes

_____

_____

_____

Teacher's Name _____ Class _____ Date _____ Room _____

Goals
    1. Use trigonometric relationships to evaluate the trigonometric functions of acute angles.
    2. Use trigonometric ratios to solve real-life problems.

State/Local Objectives _____

_____

NCTM Curriculum Standards: Problem Solving, Communication, Reasoning, Connections, Functions, Trigonometry

✓ **Check items you wish to use for this lesson.**

Introducing the Lesson
____ Teaching Tools: Problem of the Day copymasters page 59, or Teacher's Edition page 678
____ Teaching Tools: Warm-Up Exercises copymasters page 95, or Teacher's Edition page 678

    Notes for substitute teacher _____

_____

_____

Closing the Lesson
____ Communicating about Algebra, Student's Edition page 681
____ Extend Communicating, Teacher's Edition page 681
____ Guided Practice Exercises, Student's Edition page 682

Homework Assignment, pages 682–684
____ Basic/Average: Ex. 7–19 odd, 25–35 odd, 45–55 odd, 57, 61–73
____ Above Average: Ex. 9–21 odd, 27–37 odd, 47–59 odd, 64–73
____ Advanced: Ex. 9–21 odd, 27–37 odd, 43–59 odd, 64–73

Reteaching the Lesson
____ Extra Practice Copymasters page 81
____ Reteaching Copymasters page 81
____ Alternative Assessment: Math Log pages 66–68

Extending the Lesson
____ Technology, Teacher's Edition page 682
____ Enrichment, Teacher's Edition page 684
____ Research Project: Alternative Assessment page 27

Notes

_____

_____

_____

Teacher's Name _____ *Class* _____ *Date* _____ *Room* _____

Goals
    1. Measure angles in standard position using degree measure and radian measure.
    2. Use radian measure in real-life problems.

State/Local Objectives _____

_____

NCTM Curriculum Standards: Problem Solving, Communication, Connections, Geometry, Trigonometry, Structures

✓ **Check items you wish to use for this lesson.**

Introducing the Lesson
____ Teaching Tools: Problem of the Day copymasters page 59, or Teacher's Edition page 685
____ Teaching Tools: Warm-Up Exercises copymasters page 95, or Teacher's Edition page 685

Teaching the Lesson using the following:
____ Teaching Tools: transparencies pages 6, 25

    Notes for substitute teacher _____

_____

_____

Closing the Lesson
____ Communicating about Algebra, Student's Edition page 688
____ Extend Communicating: Cooperative Learning, Teacher's Edition page 688
____ Guided Practice Exercises, Student's Edition page 689

Homework Assignment, pages 689–691
____ Basic/Average: Ex. 9–19 odd, 27–53 odd, 55–62, 63, 65, 67–70
____ Above Average: Ex. 9–19 odd, 25–37 odd, 43–51 odd, 55–62, 63, 65, 67–71
____ Advanced: Ex. 11–21 odd, 25–37 odd, 43–51 odd, 55–62, 63, 65, 67–72

Reteaching the Lesson
____ Extra Practice Copymasters page 82
____ Reteaching Copymasters page 82

Extending the Lesson
____ Enrichment, Teacher's Edition page 692
____ Technology: Using Calculators and Computers page 91

Notes

_____

_____

_____

Teacher's Name_____ Class_____ Date_____ Room_____

Goals
    1. Evaluate trigonometric functions of any angle.
    2. Use trigonometric functions to solve real-life problems.

State/Local Objectives _____

_____

NCTM Curriculum Standards: Problem Solving, Communication, Reasoning, Connections, Functions, Trigonometry

✓ **Check items you wish to use for this lesson.**

Introducing the Lesson
____ Teaching Tools: Problem of the Day copymasters page 59, or Teacher's Edition page 693
____ Teaching Tools: Warm-Up Exercises copymasters page 96, or Teacher's Edition page 693

Teaching the Lesson using the following:
____ Color Transparencies pages 67, 68, 69
____ Teaching Tools: transparencies pages 2, 3, 6, 25

    Notes for substitute teacher _____

_____

_____

Closing the Lesson
____ Communicating about Algebra, Student's Edition page 696
____ Extend Communicating: Cooperative Learning, Teacher's Edition page 696
____ Guided Practice Exercises, Student's Edition page 697

Homework Assignment, pages 697–699
____ Basic/Average: Ex. 11–21 odd, 27–37 odd, 41–71 odd, 77–83 odd
____ Above Average: Ex. 11–21 odd, 25–37 odd, 42–66 multiples of 3, 70–74, 77–83 odd
____ Advanced: Ex. 11–21 odd, 27–37 odd, 39–66 multiples of 3, 70–76, 83

Reteaching the Lesson
____ Extra Practice Copymasters page 83
____ Reteaching Copymasters page 83

Extending the Lesson
____ Writing, Teacher's Edition page 697
____ Enrichment, Teacher's Edition page 698
____ Applications Handbook, pages 3, 46–48
____ Alternative Assessment page 27
____ Cultural Diversity Extensions page 39
____ Technology: Using Calculators and Computers page 93

Notes

_____

_____

_____

Teacher's Name_____ Class_____ Date_____ Room_____

Goals
    1. Evaluate inverse trigonometric functions.
    2. Use inverse trigonometric functions to solve real-life problems.

State/Local Objectives _____

_____

NCTM Curriculum Standards: Problem Solving, Communication, Connections, Functions, Trigonometry

✓ **Check items you wish to use for this lesson.**

Introducing the Lesson
____ Teaching Tools: Problem of the Day copymasters page 60, or Teacher's Edition page 703
____ Teaching Tools: Warm-Up Exercises copymasters page 96, or Teacher's Edition page 703

Teaching the Lesson using the following:
____ Technology for Example 2, Teacher's Edition page 704
____ Color Transparencies page 70

    Notes for substitute teacher _____

_____

_____

Closing the Lesson
____ Communicating about Algebra, Student's Edition page 705
____ Extend Communicating: Cooperative Learning, Teacher's Edition page 706
____ Guided Practice Exercises, Student's Edition page 706

Homework Assignment, pages 706–708
____ Basic/Average: Ex. 7–19 odd, 31–37 odd, 40, 41, 45–48
____ Above Average: Ex. 7–19 odd, 27–37 odd, 39–42, 45–50
____ Advanced: Ex. 7–19 odd, 27–37 odd, 39–45, 47, 49–52

Reteaching the Lesson
____ Extra Practice Copymasters page 84
____ Reteaching Copymasters page 84

Extending the Lesson
____ Enrichment, Teacher's Edition page 708
____ Applications Handbook, pages 38, 39

Notes

_____

_____

_____

Teacher's Name_____ Class_____ Date_____ Room_____

Goals
    1. Use the law of sines to find the sides and angles of any triangle.
    2. Use the law of sines to solve real-life problems.

State/Local Objectives _____

_____

NCTM Curriculum Standards: Problem Solving, Communication, Connections, Geometry (synthetic), Trigonometry

✓ **Check items you wish to use for this lesson.**

Introducing the Lesson
___ Teaching Tools: Problem of the Day copymasters page 60, or Teacher's Edition page 709
___ Teaching Tools: Warm-Up Exercises copymasters page 96, or Teacher's Edition page 709

Teaching the Lesson using the following:
___ Color Transparencies page 71

Notes for substitute teacher _____

_____

_____

Closing the Lesson
___ Communicating about Algebra, Student's Edition page 711
___ Guided Practice Exercises, Student's Edition page 712

Homework Assignment, pages 712–715
___ Basic/Average: Ex. 13–25 odd, 43–53 odd, 57, 61, 62, 63–71 odd, 72
___ Above Average: Ex. 9–23 odd, 29–39 odd, 45–59 odd, 62–64, 69–72
___ Advanced: Ex. 9–23 odd, 29–39 odd, 45–57 odd, 58–64, 71–72

Reteaching the Lesson
___ Extra Practice Copymasters page 85
___ Reteaching Copymasters page 85

Extending the Lesson
___ Writing, Teacher's Edition page 713
___ Research, Teacher's Edition page 714
___ Technology, Teacher's Edition page 715

Notes

_____

_____

_____

Teacher's Name _____ *Class* _____ *Date* _____ *Room* _____

Goals
    1. Use the Law of Cosines to find the sides and angles of any triangle.
    2. Use the Law of Cosines to solve real-life problems.

State/Local Objectives _____

_____

NCTM Curriculum Standards: Problem Solving, Communication, Connections, Geometry, Trigonometry

✓ **Check items you wish to use for this lesson.**

Introducing the Lesson
____ Teaching Tools: Problem of the Day copymasters page 60, or Teacher's Edition page 716
____ Teaching Tools: Warm-Up Exercises copymasters page 97, or Teacher's Edition page 716

Teaching the Lesson using the following:
____ Color Transparencies page 71

Notes for substitute teacher _____

_____

_____

Closing the Lesson
____ Communicating about Algebra, Student's Edition page 718
____ Guided Practice Exercises, Student's Edition page 719

Homework Assignment, pages 719–721
____ Basic/Average: Ex. 7–23 odd, 33–41 odd, 44, 47, 48, 51, 55
____ Above Average: Ex. 7–23 odd, 31–39 odd, 43–48, 55, 56
____ Advanced: Ex. 7–23 odd, 29–39 odd, 43–48, 56, 57

Reteaching the Lesson
____ Extra Practice Copymasters page 86
____ Reteaching Copymasters page 86

Extending the Lesson
____ Project, Teacher's Edition page 720
____ Enrichment, Teacher's Edition page 721
____ Applications Handbook page 38

Notes

_____

_____

_____

Teacher's Name _____ *Class* _____ *Date* _____ *Room* _____

Goals
1. Sketch the graphs of the sine and cosine functions.
2. Use the sine and cosine functions as models of real-life problems.

State/Local Objectives _____

_____

NCTM Curriculum Standards: Problem Solving, Communication, Connections, Functions, Trigonometry

✓ **Check items you wish to use for this lesson.**

Introducing the Lesson
____ Teaching Tools: Problem of the Day copymasters page 61, or Teacher's Edition page 730
____ Teaching Tools: Warm-Up Exercises copymasters page 97, or Teacher's Edition page 730

Teaching the Lesson using the following:
____ Color Transparencies page 73
____ Teaching Tools: transparencies pages 26, 27, 28

Notes for substitute teacher _____

_____

_____

Closing the Lesson
____ Communicating about Algebra, Student's Edition page 733
____ Extend Communicating: Cooperative Learning, Teacher's Edition page 733
____ Guided Practice Exercises, Student's Edition page 734

Homework Assignment, pages 734–736
____ Basic/Average: Ex. 7, 8, 10, 12, 16–20, 25–41 odd, 44, 48–51
____ Above Average: Ex. 7, 8, 10, 12, 16–20, 25–41 odd, 42, 43–52
____ Advanced: Ex. 7, 8, 10, 12, 16–20, 25–41 odd, 42–47, 50–52

Reteaching the Lesson
____ Extra Practice Copymasters page 87
____ Reteaching Copymasters page 87
____ Alternative Assessment: Math Log pages 69–71

Extending the Lesson
____ Writing, Teacher's Edition page 734
____ Enrichment, Teacher's Edition pages 735–736
____ Applications Handbook pages 49–51, 56, 57
____ Cooperative Learning, Alternative Assessment page 28
____ Cultural Diversity Extensions page 40
____ Technology: Using Calculators and Computers page 95

Notes

_____

_____

_____

Omit days for Basic Course, 2 days for Full Course

Teacher's Name_____ Class_____ Date_____ Room_____

Goals   1. Graph vertical and horizontal shifts and reflections of the graphs of the sine and cosine functions.
2. Use sine and cosine functions to solve real-life problems.

State/Local Objectives _____

_____

NCTM Curriculum Standards: Problem Solving, Communication, Reasoning, Connections, Functions, Geometry (algebraic), Trigonometry

✓ **Check items you wish to use for this lesson.**

Introducing the Lesson
____ Teaching Tools: Problem of the Day copymasters page 61, or Teacher's Edition page 737
____ Teaching Tools: Warm-Up Exercises copymasters page 98, or Teacher's Edition page 737

Teaching the Lesson using the following:
____ Color Transparencies page 74
____ Teaching Tools: transparencies pages 26, 27, 28

Notes for substitute teacher _____

_____

_____

Closing the Lesson
____ Communicating about Algebra, Student's Edition page 739
____ Guided Practice Exercises, Student's Edition page 740

Homework Assignment, pages 740–742
____ Basic/Average: Ex. 11–17 odd, 19–24, 31–39 odd, 45–51, 59–62
____ Above Average: Ex. 9–19 odd, 20–24, 29–39 odd, 45–51, 59–64
____ Advanced: Ex. 9–19 odd, 20–24, 29–39 odd, 45–54, 61–64

Reteaching the Lesson
____ Extra Practice Copymasters page 88
____ Reteaching Copymasters page 88

Extending the Lesson
____ Project, Teacher's Edition page 741
____ Technology, Teacher's Edition pages 741, 742
____ Writing, Teacher's Edition page 741
____ Enrichment, Teacher's Edition page 742
____ Applications Handbook page 3
____ Technology: Using Calculators and Computers page 97

Notes

_____

_____

ⓒ D. C. Heath and Company

Teacher's Name _____ *Class* _____ *Date* _____ *Room* _____

Goals
1. Use trigonometric identities to simplify trigonometric expressions and verify trigonometric identities.
2. Use trigonometric identities to solve real-life problems.

State/Local Objectives _____

_____

NCTM Curriculum Standards: Problem Solving, Communication, Connections, Functions, Trigonometry

✓ **Check items you wish to use for this lesson.**

Introducing the Lesson
___ Teaching Tools: Problem of the Day copymasters page 61, or Teacher's Edition page 744
___ Teaching Tools: Warm-Up Exercises copymasters page 98, or Teacher's Edition page 744

Teaching the Lesson using the following:
___ Technology for Example 6, Teacher's Edition page 745
___ Color Transparencies page 74
___ Teaching Tools: transparencies pages 2, 3

Notes for substitute teacher _____

_____

_____

Closing the Lesson
___ Communicating about Algebra, Student's Edition page 746
___ Guided Practice Exercises, Student's Edition page 747

Homework Assignment, pages 747–749
___ Basic/Average: Ex. 9–17 odd, 27–35 odd, 41, 45, 49–55 odd
___ Above Average: Ex. 9–19 odd, 25–35 odd, 39–45 odd, 46, 54–57
___ Advanced: Ex. 9–19 odd, 25–35 odd, 39–43 odd, 44–47, 56–58

Reteaching the Lesson
___ Extra Practice Copymasters page 89
___ Reteaching Copymasters page 89

Extending the Lesson
___ Technology, Teacher's Edition page 748
___ Enrichment, Teacher's Edition page 749
___ Cooperative Learning, Alternative Assessment page 28
___ Technology, Alternative Assessment page 28
___ Technology: Using Calculators and Computers page 99

Notes

_____

_____

_____

Teacher's Name _____ Class _____ Date _____ Room _____

Goals   1. Solve trigonometric equations.
2. Solve trigonometric equations that model real-life problems.

State/Local Objectives _____

_____

NCTM Curriculum Standards: Problem Solving, Communication, Reasoning, Connections, Functions,
Geometry (algebraic), Trigonometry

✓ **Check items you wish to use for this lesson.**

Introducing the Lesson
___ Teaching Tools: Problem of the Day copymasters page 62, or Teacher's Edition page 753
___ Teaching Tools: Warm-Up Exercises copymasters page 98, or Teacher's Edition page 753

Teaching the Lesson using the following:
___ Color Transparencies pages 75, 76

Notes for substitute teacher _____

_____

_____

Closing the Lesson
___ Communicating about Algebra, Student's Edition page 756
___ Extend Communicating, Teacher's Edition page 756
___ Guided Practice Exercises, Student's Edition page 757

Homework Assignment, pages 757–759
___ Basic/Average: Ex. 11–25 odd, 39–45 odd, 52–54, 55–67 odd, 68
___ Above Average: Ex. 9–27 odd, 33–45 odd, 49–54, 65–69
___ Advanced: Ex. 9–27 odd, 33–45 odd, 49–54, 67–70

Reteaching the Lesson
___ Extra Practice Copymasters page 90
___ Reteaching Copymasters page 90

Extending the Lesson
___ Technology, Teacher's Edition page 757
___ Technology, Teacher's Edition page 758
___ Enrichment, Teacher's Edition page 759

Notes

_____

_____

_____

ⓒ D. C. Heath and Company

Teacher's Name _____ Class _____ Date _____ Room _____

Goals
1. Use the sum and difference formulas to evaluate trigonometric functions of the sum and difference of two angles.
2. Use sum and difference formulas to solve real-life problems.

State/Local Objectives _____

_____

NCTM Curriculum Standards: Problem Solving, Communication, Reasoning, Connections, Functions, Trigonometry

✓ **Check items you wish to use for this lesson.**

Introducing the Lesson
___ Teaching Tools: Problem of the Day copymasters page 62, or Teacher's Edition page 760
___ Teaching Tools: Warm-Up Exercises copymasters page 99, or Teacher's Edition page 760

Teaching the Lesson using the following:
___ Color Transparencies pages 76, 77

Notes for substitute teacher _____

_____

_____

Closing the Lesson
___ Communicating about Algebra, Student's Edition page 762
___ Extend Communicating: Cooperative Learning, Teacher's Edition page 762
___ Guided Practice Exercises, Student's Edition page 763

Homework Assignment, pages 763–765
___ Basic/Average: Ex. 9–17 odd, 21–29 odd, 33, 37, 38, 41–44, 49, 50
___ Above Average: Ex. 7–17 odd, 19–29 odd, 35, 37–40, 44, 49, 51
___ Advanced: Ex. 7–17 odd, 19–29 odd, 31, 35, 37–40, 43, 47, 51

Reteaching the Lesson
___ Extra Practice Copymasters page 91
___ Reteaching Copymasters page 91

Extending the Lesson
___ Enrichment, Teacher's Edition page 765
___ Applications Handbook pages 38, 39, 56, 57

Notes

_____

_____

_____

Teacher's Name_____ Class_____ Date_____ Room_____

Goals   1. Use double-angle and half-angle formulas.
               2. Use trigonometry to solve real-life problems.

State/Local Objectives _____

_____

NCTM Curriculum Standards: Problem Solving, Communication, Connections, Functions, Geometry (algebraic),
                                Trigonometry

✓ **Check items you wish to use for this lesson.**

Introducing the Lesson
____ Teaching Tools: Problem of the Day copymasters page 62, or Teacher's Edition page 766
____ Teaching Tools: Warm-Up Exercises copymasters page 99, or Teacher's Edition page 766

    Notes for substitute teacher _____

_____

_____

Closing the Lesson
____ Communicating about Algebra, Student's Edition page 768
____ Extend Communicating: Cooperative Learning, Teacher's Edition page 768
____ Guided Practice Exercises, Student's Edition page 769

Homework Assignment, pages 769–771
____ Basic/Average: Ex. 7–19 odd, 25–31 odd, 36–40 even, 41–43, 46, 48–50
____ Above Average: Ex. 7–21 odd, 27–33 odd, 36–40 even, 41–46, 48–50
____ Advanced: Ex. 7–21 odd, 27–33 odd, 38–46, 47, 51, 52

Reteaching the Lesson
____ Extra Practice Copymasters page 92
____ Reteaching Copymasters page 92

Extending the Lesson
____ Technology, Teacher's Edition page 769
____ Enrichment, Teacher's Edition page 771
____ Applications Handbook pages 55, 56

Notes

_____

_____

_____

Teacher's Name _____ Class _____ Date _____ Room _____

Goals
1. Find the probability of an event.
2. Use probability to answer questions about real life.

State/Local Objectives _____

_____

NCTM Curriculum Standards: Problem Solving, Communication, Connections, Probability, Discrete Math

✓ **Check items you wish to use for this lesson.**

Introducing the Lesson
___ Teaching Tools: Problem of the Day copymasters page 63, or Teacher's Edition page 782
___ Teaching Tools: Warm-Up Exercises copymasters page 99, or Teacher's Edition page 782

Teaching the Lesson using the following:
___ Cooperative Learning: Example 4, Teacher's Edition page 784
___ Extension for Example 2, Teacher's Edition page 784
___ Color Transparencies page 79

Notes for substitute teacher _____

_____

_____

Closing the Lesson
___ Communicating about Algebra, Student's Edition page 784
___ Guided Practice Exercises, Student's Edition page 785

Homework Assignment, pages 785–787
___ Basic/Average: Ex. 7, 10–21, 22–25, 28–30
___ Above Average: Ex. 8–23, 26–30
___ Advanced: Ex. 7–23, 25–30

Reteaching the Lesson
___ Extra Practice Copymasters page 93
___ Reteaching Copymasters page 93
___ Alternative Assessment: Math Log pages 72–74

Extending the Lesson
___ Enrichment, Teacher's Edition page 787
___ Applications Handbook pages 13, 14
___ Cultural Diversity Extensions page 41
___ Research Project: Alternative Assessment pages 28–29

Notes

_____

_____

_____

Omit days for Basic Course, 2 days for Full Course

Teacher's Name _____ Class _____ Date _____ Room _____

Goals
1. Use the Fundamental Counting Principle to find probabilities.
2. Use permutations to find probabilities.

State/Local Objectives _____

_____

NCTM Curriculum Standards: Problem Solving, Communication, Reasoning, Connections, Probability,
Discrete Math

✓ **Check items you wish to use for this lesson.**

Introducing the Lesson
___ Teaching Tools: Problem of the Day copymasters page 63, or Teacher's Edition page 788
___ Teaching Tools: Warm-Up Exercises copymasters page 100, or Teacher's Edition page 788

Teaching the Lesson using the following:
___ Extension for Example 3, Teacher's Edition page 789
___ Color Transparencies page 80

Notes for substitute teacher _____

_____

_____

Closing the Lesson
___ Communicating about Algebra, Student's Edition page 791
___ Guided Practice Exercises, Student's Edition page 792

Homework Assignment, pages 792–794
___ Basic/Average: Ex. 7, 11, 14, 15, 16, 17, 23, 24, 26–34
___ Above Average: Ex. 9–13 odd, 14–20, 23–26, 32–34
___ Advanced: Ex. 9–13 odd, 14–26, 32–34

Reteaching the Lesson
___ Extra Practice Copymasters page 94
___ Reteaching Copymasters page 94

Extending the Lesson
___ Enrichment, Teacher's Edition page 794
___ Cooperative Learning, Alternative Assessment page 29

Notes

_____

_____

_____

© D.C. Heath and Company

Teacher's Name _____ *Class* _____ *Date* _____ *Room* _____

Goals
1. Use combinations to count the number of ways an event can happen.
2. Use combinations to find probabilities.

State/Local Objectives _____

_____

NCTM Curriculum Standards: Problem Solving, Communication, Reasoning, Connections, Probability, Discrete Math

✓ **Check items you wish to use for this lesson.**

Introducing the Lesson
___ Teaching Tools: Problem of the Day copymasters page 63, or Teacher's Edition page 796
___ Teaching Tools: Warm-Up Exercises copymasters page 100, or Teacher's Edition page 796

Teaching the Lesson using the following:
___ Teaching Tools: transparencies pages 2, 3, 24

Notes for substitute teacher _____

_____

_____

Closing the Lesson
___ Communicating about Algebra, Student's Edition page 798
___ Guided Practice Exercises, Student's Edition page 799

Homework Assignment, pages 799–801
___ Basic/Average: Ex. 5, 8, 9, 12, 14, 15, 21–28
___ Above Average: Ex. 5–9, 12, 15–22, 27–29
___ Advanced: Ex. 5–9, 12, 14–21, 27–30

Reaching the Lesson
___ Extra Practice Copymasters page 95
___ Reteaching Copymasters page 95

Extending the Lesson
___ Enrichment, Teacher's Edition page 801
___ Technology: Using Calculators and Computers page 101

Notes

_____

_____

_____

Omit days for Basic Course, 2 days for Full Course

Teacher's Name_____ Class_____ Date_____ Room_____

Goals
1. Use unions to find probabilities.
2. Use complements and intersections to find probabilities.

State/Local Objectives _____

_____

NCTM Curriculum Standards: Problem Solving, Communication, Reasoning, Connections, Probability, Discrete Math

✓ **Check items you wish to use for this lesson.**

Introducing the Lesson
____ Teaching Tools: Problem of the Day copymasters page 64, or Teacher's Edition page 805
____ Teaching Tools: Warm-Up Exercises copymasters page 101, or Teacher's Edition page 805

Teaching the Lesson using the following:
____ Extension for Example 2, Teacher's Edition page 806
____ Color Transparencies page 81

Notes for substitute teacher _____

_____

_____

Closing the Lesson
____ Communicating about Algebra, Student's Edition page 807
____ Guided Practice Exercises, Student's Edition page 808

Homework Assignment, pages 808–810
____ Basic/Average: Ex. 7–15 odd, 18, 20, 21, 23, 24, 27–31
____ Above Average: Ex. 7–15 odd, 17–23 odd, 24–31
____ Advanced: Ex. 7–15 odd, 18–24, 26–31

Reteaching the Lesson
____ Extra Practice Copymasters page 96
____ Reteaching Copymasters page 96

Extending the Lesson
____ Enrichment, Teacher's Edition page 810
____ Technology: Using Calculators and Computers page 103

Notes

_____

_____

_____

Ⓒ D. C. Heath and Company

Teacher's Name_____ Class_____ Date_____ Room_____

Goals   1. Find the probability of independent events.
2. Use the complement of the event to find the probability of the event.

State/Local Objectives _____

_____

NCTM Curriculum Standards: Problem Solving, Communication, Reasoning, Connections, Probability,
Discrete Math

✓ **Check items you wish to use for this lesson.**

Introducing the Lesson
____ Teaching Tools: Problem of the Day copymasters page 64, or Teacher's Edition page 811
____ Teaching Tools: Warm-Up Exercises copymasters page 101, or Teacher's Edition page 811

Teaching the Lesson using the following:
____ Technology for Example 1, Teacher's Edition page 811

Notes for substitute teacher _____

_____

_____

Closing the Lesson
____ Communicating about Algebra, Student's Edition page 813
____ Extend Communicating: Binomial Trials, Teacher's Edition pages 813–814
____ Guided Practice Exercises, Student's Edition page 814

Homework Assignment, pages 814–816
____ Basic/Average: Ex. 7–13 odd, 16–19, 21–23
____ Above Average: Ex. 7–15 odd, 16–22, 24
____ Advanced: Ex. 7–15 odd, 16–24

Reteaching the Lesson
____ Extra Practice Copymasters page 97
____ Reteaching Copymasters page 97

Extending the Lesson
____ Research, Teacher's Edition page 815
____ Enrichment, Teacher's Edition page 816

Notes

_____

_____

_____

Omit days for Basic Course, 2 days for Full Course

Teacher's Name _____ Class _____ Date _____ Room _____

Goals   1. Find the expected value of a sample space.
2. Use expected value to answer questions about real-life situations.

State/Local Objectives _____

_____

NCTM Curriculum Standards: Problem Solving, Communication, Reasoning, Connections, Probability,
Discrete Math

✓ **Check items you wish to use for this lesson.**

Introducing the Lesson
____ Teaching Tools: Problem of the Day copymasters page 64, or Teacher's Edition page 817
____ Teaching Tools: Warm-Up Exercises copymasters page 101, or Teacher's Edition page 817

Notes for substitute teacher _____

_____

_____

Closing the Lesson
____ Communicating about Algebra, Student's Edition page 819
____ Extend Communicating, Teacher's Edition page 820
____ Guided Practice Exercises, Student's Edition page 820

Homework Assignment, pages 820–822
____ Basic/Average: Ex. 5–11, 13–15, 18–21
____ Above Average: Ex. 5–15, 17–21
____ Advanced: Ex. 5–21

Reteaching the Lesson
____ Extra Practice Copymasters page 98
____ Reteaching Copymasters page 98

Extending the Lesson
____ Enrichment, Teacher's Edition page 822
____ Cooperative Learning, Alternative Assessment page 29

Notes

_____

_____

_____